RESTRICTED

Army Code No. 71711

BRITISH ARMY

MILITARY TRACKING

2001

RESTRICTED

RESTRICTED Army Code No. 71711

Military Tracking

Prepared under the direction

of the Chief of the General Staff

Ministry of Defence

August 2001

RESTRICTED **2001**

The Infantry Mission is —

'to close with and engage the enemy in concert with other Arms, in all operational theatres and environments, in order to bring about his defeat'.

RESTRICTED

FOREWORD

1. The development of doctrine is a continuous process and the information contained in the series of Infantry Doctrine Publications will naturally be subject to change. This change will either be driven from higher level developments or from field experience. Any person wishing to propose amendments to the pamphlets in Volume 1 is strongly encouraged to write to SO3 TD, Force Development Branch, Headquarters Infantry, Warminster Training Centre, Warminster, Wiltshire, BA12 0DJ. All proposals will be considered for submission to the Infantry Doctrine Working Group. Other queries should be made to Infantry Publications (same address), ATN Warminster Mil ext 2325/2452.

2. Infantry Doctrine forms part of a coherent hierarchy of doctrine publications. Associated publications are: British Defence Doctrine; The British Military Doctrine; ADP Volume 1 – Operations; ADP Volume 2 – Command AFM Volume 1 Part 2 – Battlegroup Tactics; Army Unit SOPs, and the Tactical Aide-Memoire. British Army Review, Army Training and Doctrine News and The Infantryman provide current thought and discussion on doctrine matters.

3. Individuals should first read the Introduction on page xi which explains the aim, layout and scope of this pamphlet as well as the structure and breakdown of Infantry Tactical Doctrine Publications.

DISTRIBUTION (1)

HQs Commands, Divisions, Brigades and Directorates (less Infantry)	One copy each
HQ TSC(L)	2 copies
HQ TSC(G)	2 copies
DGD&D — LW1	2 copies
Infantry	
HQ Infantry/Infantry Reps and LOs	10 copies
Regular Battalions	20 copies each
TA Battalions	One copy each
Independent/Detached Infantry Companies	2 copies each
Jungle Warfare Wing – Brunei	100 copies
ITDU	One copy
SAS Regiments	20 copies each
HQ School of Infantry	2 copies
HQ Infantry Battle School	5 copies
ITC Warminster	5 copies
ITC Wales	5 copies
ITC Catterick	5 copies
Army Training Regiments/Training Depots	One copy each

RESTRICTED

v

RESTRICTED

Royal Military Academy Sandhurst	2 copies
HQ LWTC	2 copies
HQ LWS	2 copies
HQ LWCTG	2 copies
BATSUB	5 copies
BATLSK	5 copies
JSCSC	20 copies
AJD (JSCSC)	20 copies
Defence Animal Centre	5 copies
OPTAG (K)	5 copies
Royal Marines	
HQ Royal Marines	10 copies
HQ 3 Commando Brigade	2 copies
Commandos	20 copies each
CTC RM	2 copies
RAF Regiment Squadrons	2 copies each

Notes:

1. *General.*

    a. The previous system of A to F scaling is being phased out for GSTPs as new or revised publications are issued. Units will in future receive a specific initial entitlement as decided by the sponsor.
    b. Requests for additional copies must be fully justified and are to be made to DGD&D, Publications in the first instance. Requests will fall into one of three categories and are to be specified:
      (1) *Replacement Issue.* To bring units back up to entitlement as a result of loss or damage.
      (2) *Supplementary Issue.* This applies when a unit's role or circumstance changes necessitating as increase to the original entitlement.
      (3) *New Issue.* This is where a unit has not previously been scaled for a particular publication.
    c. Units are reminded that all publications are accountable documents and their issue and receipt should be the responsibility of a nominated officer or SNCO.

RESTRICTED

CONTENTS

	Page
Foreword	v
Distribution	v
Contents	vii
List of Illustrations	viii
Introduction	xi
Annex A. — Basic Tracking Terminology	xv

CHAPTER 1. — INTRODUCTION TO TRACKING

Section 1. — General	1-1
Section 2. — Types of Tracking	1-5
Section 3. — Qualities of a Tracker	1-6

CHAPTER 2. — ELEMENTS OF TRACKING (EOT)

Section 1. — General	2-1
Section 2. — EOT 1 — Sign	2-1
Section 3. — EOT 2 — Factors Which Affect Tracking	2-4
Section 4. — EOT 3 — Judging the Age of Sign	2-10
Section 5. — EOT 4 — Information Gained From Tracking	2-13
Annex A. — Observation and the Use of Other Senses	2A-1
Annex B. — Methods of Interpretation and Assumption	2B-1

CHAPTER 3. — TRACKING TECHNIQUES AND PROCEDURES

Section 1. — General	3-1
Section 2. — Track Casting Drill	3-1
Section 3. — Track Pursuit Drill	3-5
Section 4. — Duties of a Coverman	3-8
Section 5. — Incident Tracking Drill	3-10
Section 6. — Track Isolation Drill	3-13
Section 7. — Deception Tactics	3-15

CHAPTER 4. — TACTICAL TRACKING

Section 1. — General	4-1
Section 2. — Tracker Team Composition	4-1

RESTRICTED

		Page
Section 3.	— Duties of a Lead Scout	4-5
Section 4.	— Tracker Team Routine Drills	4-7
Section 5.	— The Pursuit	4-12
Annex A	— Tracker Team Kit and Equipment	4A-1

CHAPTER 5. — TRACKING REPORTING

Section 1.	— General	5-1
Section 2.	— Signals Reports	5-2
Section 3.	— Hot Debrief	5-3
Section 4.	— Visual Tracking Report	5-7
Section 5.	— Tracking Presentation	5-15
Annex A	— Signals Report Format	5A-1
Annex B	— Hot Debrief Report Format	5B-1
Annex C	— Visual Tracking Patrol Report Format	5C-1

CHAPTER 6. — TRAINING THE TRACKER

Section 1.	— General	6-1
Section 2.	— Training the Visual Tracker	6-1
Section 3.	— Training the Tracker Team	6-7
Section 4.	— Track and Incident Laying	6-10
Section 5.	— Tracking Presentations	6-11
Annex A	— Memory Training	6A-1
Annex B	— Track Laying Report	6B-1

ILLUSTRATIONS

Figure No   **Page**

1	Track Casting Drill — The Initial Probe	3-2
2	Track Casting Drill — The Initial Cast	3-3
3	Track Casting Drill — The Extended Cast	3-4
4	Track Pursuit Drill — Step 1	3-6
5	Track Pursuit Drill — Step 3	3-7
6	Tracker Team Order of March	4-8
7	Cold Track Formation	4-14
8	Warm Track Formation	4-15

RESTRICTED

RESTRICTED

		Page
9	Hot Track Formation	4-16
10	Formation With a Tracker Dog Leading	4-17
11	An Example of a Completed Patrol Details Box	5-8
12	An Example of a Completed Cast Site Details Box	5-9
13	An Example of a Route Followed	5-10
14	An Example of an Incident Encountered	5-14
15	Pace to Pace Tracking Box	6-3

RESTRICTED

RESTRICTED

INFANTRY TACTICAL DOCTRINE

VOLUME 1 — THE INFANTRY COMPANY GROUP

PAMPHLET NO. 6

MILITARY TRACKING

INTRODUCTION

Aim

1. The aim of Pamphlet No. 6 is to provide the necessary information and tactical skills needed in order to train and use Military Tracking Teams effectively in the jungle and other environments.

CONTENTS
CHAPTER 1. — INTRODUCTION TO TRACKING
CHAPTER 2. — ELEMENTS OF TRACKING (EOT)
CHAPTER 3. — TRACKING TECHNIQUES AND PROCEDURES
CHAPTER 4. — TACTICAL TRACKING
CHAPTER 5. — TRACKING REPORTING
CHAPTER 6. — TRAINING THE TRACKER

Layout

2. This pamphlet is laid out in six chapters (*see* Contents). Although not presented in lesson form the contents of each chapter cover the instruction given on the courses run in Brunei by the Jungle Warfare Wing. A list of the basic tracking terminology used in this pamphlet is at Annex A.

Scope

3. Infantry Tactical Doctrine Volume 1 The Infantry Company Group Pamphlet No. 5 Infantry Company Group Jungle Tactics, deals in depth with the conditions found in the jungle environment and covers the tactics, techniques and procedures for operating in it. It also covers all aspects of training in the jungle environment. This pamphlet is designed to be used by both military tracking instructors and military trackers who should all be fully capable of operating in the jungle for a prolonged period of time. As these skills and techniques may be adapted for use in any theatre this pamphlet should be read in conjunction with the appropriate Tactical Doctrine Pamphlet, particularly Pamphlet No. 3, Infantry Platoon Tactics.

RESTRICTED

Training

4. Chapter 6 comprehensively covers all aspects of instructor and unit training. The Jungle Warfare Wing currently runs the following courses per year:

   a. *The Jungle Warfare Instructor's Course (JWIC).* This is an eight week course run three times a year.

   b. *The Jungle Warfare Tracking Instructor's Course (JWTIC).* This is a five week course run twice a year.

Structure of Infantry Tactical Doctrine Publications

5. ***The Doctrine Hierarchy.*** The Doctrine Hierarchy ensures that Army doctrine cascades from policy and principles to practical applications and procedures. There are three elements to the hierarchy:

   a. *Principles.* Principles are the guiding doctrine that is concerned with operational art and the British outlook on operations is contained in British Defence Doctrine (JWP 0–01). From this overarching publication is derived British Military Doctrine (BMD), Army Doctrine Publications (ADPs) and various Joint publications. Allied publications, such as ATP–35(B), also describe principles.

   b. *Practices.* The practical application of tactical doctrine for field commanders at formation and battlegroup level are contained in Army Field Manuals (AFMs). The Infantry tactical doctrine contained in this new series of pamphlets lies between practices and procedures.

   c. *Procedures.* Procedures are the mechanics that ensure the success of operations. At Army level procedures are detailed in publications such as AFSOPs, AUSOPs and TAM. Infantry procedures are detailed in a new range of pamphlets called Infantry Tactical Doctrine Publications.

6. ***Infantry Doctrine Development.*** The Director of Infantry is charged by the Chief of the General Staff with the development of Infantry tactical doctrine at company level and below. Development of this doctrine is the responsibility of the Infantry Force Development Committee (IFDC) whose aim is to integrate the five functional areas of Infantry Fighting Power: doctrine, equipment, structures, training and manpower and logistics. Each of these functional areas is represented by a Working Group whose aim is to provide advice and guidance to the IFDC in their area, in conjunction with the Force Development branch of Headquarters Infantry.

7. ***Infantry Tactical Doctrine.*** Infantry tactical doctrine has been defined as 'tactical doctrine to provide guidance to company commanders by detailing Tactics, Techniques and Procedures (TTPs) appropriate to their level of tactical command'.

RESTRICTED

8. **Publications.** The structure of Infantry tactical doctrine and training publications has been rationalized and will now be broken down into two parts — Infantry Tactical Doctrine and Infantry Training — each divided into a number of volumes further sub-divided into pamphlets.

    a. *Infantry Tactical Doctrine.* Infantry tactical doctrine consists of three volumes: Volume 1 — The Infantry Company Group; Volume 2 — The Tactical Employment of Infantry Weapons and Systems; Volume 3 — Infantry Tactical Doctrine Notes.

    b. *Infantry Training Publications.* Infantry training publications consists of a number of volumes concerning skill at arms, ranges, support weapons, signals training and vehicles and vehicle mounted weapon systems.

Breakdown of Infantry Tactical Doctrine Publications

9. Infantry tactical doctrine is contained in three volumes:

    a. Volume 1 — The Infantry Company Group.

    b. Volume 2 — The Tactical Employment of Infantry Weapons and Systems.

    c. Volume 3 — Infantry Tactical Doctrine Notes.

10. The detailed breakdown of the pamphlets is as follows:

    a. *Volume 1 — The Infantry Company Group.*

Pamphlet No	Title
1	The Infantry Company Group — The Fundamentals
2	Infantry Company Group Tactics
3	Infantry Platoon Tactics
4	Armoured Infantry Company Group Tactics
5	Infantry Company Group Jungle Tactics
6	Infantry Company Group — Military Tracking

RESTRICTED

b. *Volume 2 — The Tactical Employment of Infantry Weapons and Systems.*

Pamphlet No	Title
1	RESERVED FOR FUTURE USE
2	The Medium Mortar — 81 mm L16
3	Infantry Anti-Armour Weapons: Medium Range Anti-Tank Guided Weapon — MILAN and Rocket System 94 mm HEAT — LAW

c. *Volume 3 — Infantry Tactical Doctrine Notes.*

**ANNEX A TO
INTRODUCTION**

BASIC TRACKING TERMINOLOGY

Introduction

1. Tracking is not a black art and in order for trackers and commanders to understand the information reported definitions are kept as simple as possible. Specific definitions are amplified at the relevant section however the following list covers the basic terminology:

   a. *Casting.* This is a drill used to locate or relocate sign. For example where a tracker is endeavouring to locate sign at a start point/incident area, or locate sign that has been abandoned or lost.

   b. *Cold Track.* A track which is more than one day old.

   c. *Combat Tracking/Combat Tracker Team.* Tracking at speed in order to close with the target to re-establish contact. (This is the terminology used by NZ SAS when pursuit tracking.)

   d. *Conclusive.* Sign classified as being 'conclusive' is deemed decisive sign belonging to the target.

   e. *Deception.* Deception tactics are the attempts made by the enemy to lose/delay the tracker.

   f. *Elements of Tracking (EOT).* A term used to describe the four main aspects of tracking, particularly when training trackers. (EOT 1 – 4)

   g. *Entry and Exit Points.* These are points or passage ways through which a single or number of any one quarry may pass from one environment to another.

   h. *Environment.* The immediate surroundings or conditions of life or growth. The environment will frequently change for example, moving from within the jungle, into a stream then back into the jungle, will incorporate an environment of leaves on the floor, to that of sand, then back to leaves.

   i. *Foul Track.* Is defined as a track which has entered an area where numerous other tracks exist. These may be animal tracks, vehicle tracks or the track left by another target group.

   j. *Found/Located.* Information gained through searching, for example the information identified at an incident site.

RESTRICTED

k. *Gait.* The individual's manner of walking. Identifying the gait of an enemy will assist a tracker in identifying a sign pattern amongst his quarry.

l. *Hot Debrief.* An immediate debrief on completion of a tracking task given in order to pass on timely and accurate tactical information to commanders.

m. *Hot Track.* A track which is less than two hours old.

n. *Incident.* An incident/incident site is an area where the sign indicates to the tracker that the enemy has stopped and carried out some form of action. Examples include an enemy tactical pause, tactical halt or overnight halt.

o. *Inconclusive.* Sign that is classified as being 'inconclusive' is sign that can not be confirmed as belonging to the target.

p. *Indicator Pace.* This is the pace immediately before or after a change of direction. An indicator pace is normally a shorter pace.

q. *Key Sign.* This is that sign which is prominent in any one environment within a track and serves to increase the range and speed of tracking, i.e., the sign that stands out the most to the tracker, for example in primary jungle — colour change (overturned leaves).

r. *Last Definite Sign (LDS).* The last definite sign is the last point at which the tracker can identify enemy sign. Whilst tracking the tracker should never pass his last definite sign. An exception would be when the tracker is going to probe openings. Under such circumstances the last definite sign is marked before the tracker moves beyond it.

s. *Opening.* An entrance into the undergrowth where the target may have moved through.

t. *Pace Tracking.* This is 'pace for pace' tracking carried out during the 'tuning-in' period.

u. *Patrol Report.* The report form used to record the information gained from a tracking patrol task.

v. *Patrol Signals Report.* Signals message sent at the end of each day.

w. *Pointers.* These are those signs that indicate the direction of movement, i.e., low foliage (grass) when tramped on or brushed forward, will remain in the direction of movement until disturbed again.

x. *Probe.* To investigate an area/opening when looking for further sign.

y. *Quarry/Target.* Persons who are being followed or pursued.

RESTRICTED

RESTRICTED

z. *Sign Pattern.* This is the sign that serves to indicate the habits or the peculiarities of an individual or group. For example the enemy patrol may have an individual who continually snaps twigs on route.

aa. *Sign/Spoor.* The marks inflicted by a target on the environment. For example snapped twigs, bruised roots, etc. Sign can be defined as:

> 'Any evidence of change from the natural state that is inflicted upon the environment by the passage of man, animal or machinery'.

bb. *Straight Edge.* This is the line found on leaves or blades of grass caused by the application of pressure, as in the folding of paper.

cc. *Time Bracket.* All sign in the initial stages is placed into a time bracket, i.e., the time lapse between the earliest possible time that the sign could have been made and the time it was located, in the bracket, i.e., the age of sign.

dd. *To Follow.* Proceeding after the enemy without actively seeking to catch them. Trackers follow the enemy whilst employed on information reporting tasks.

ee. *To Pursue/In Pursuit.* Actively following the enemy in order to close with them to establish contact.

ff. *Track.* The path of the target identified by the sign left.

gg. *Track Isolation.* This is where the tracker having anticipated the intended route of the target is able to 'leap frog' ahead and through the sign pattern of his target is able to identify the same track.

hh. *Track Picture.* This is the overall picture (summary of findings) gained by the tracker over any given length of track.

ii. *Tracking Presentation.* A tracking presentation is the term used to describe a formal presentation by a tracker following a patrol on the information gained from a track. It is an instructional tool used normally during training however can be used for a specific purpose during operations.

jj. *Track Pursuit Drill.* The procedure employed whilst visual tracking.

kk. *Tracker Team (TT).* This is the British Army's fundamental tracking unit.

ll. *Tuning-in.* This is the initial reading of the sign that enables the tracker to think and act as the target being pursued. Each time the tracker identifies a change in environment he should stop and tune-in. He should think what sign is going to be left in this environment, what is likely to be the key sign, and how will the target move in this environment, i.e., how will he negotiate obstacles, etc.

RESTRICTED

RESTRICTED

Tuning-in properly will assist the tracker in anticipating the enemy's future actions.

mm. *Warm Track.* A track which is between two and 24 hours old.

2. ***Summary.*** Trackers must be encouraged to use plain and simple english in order to avoid complicated explanations and to ensure that all information is passed on quickly and accurately.

RESTRICTED

Chapter 1

INTRODUCTION TO TRACKING

SECTION 1. — GENERAL

"Special training, not special men"

Introduction

0101. Visual tracking is the skill of being able to follow a man or group of men from the sign they have left. It is a very precise skill and all trackers require considerable practice if they are to achieve and maintain a high standard. There is an obvious requirement to be able to interpret sign left by an enemy as this will yield vital information. Units should endeavour to have sufficient quantities of qualified trackers when deployed to jungle theatres. This skill is not confined to the jungle environment as it can be adapted for use in any theatre.

Contents	
	Page
SECTION 1. — GENERAL	1–1
SECTION 2. — TYPES OF TRACKING	1–5
SECTION 3. — QUALITIES OF A TRACKER	1–6

0102. Tracking has been employed in numerous military campaigns. It was employed to great effect in the Malayan Emergency, the Borneo Confrontation, Vietnam and, more recently, East Timor. Countless reports of successful operations in these areas have been associated or accredited either directly or indirectly to the intelligence gathered by the trackers.

Current Tracking Applications

0103. The vast majority of commanders are unacquainted or unaware of the potential use of trained trackers. Throughout the world trackers are currently employed in the following ways:

   a. *Search and Rescue (SAR).* SAR teams can utilise tracking techniques in order to locate downed aircrew or missing personnel. However, the majority of dedicated SAR teams do not possess qualified trackers, and often SAR personnel obliterate vital sign before trackers are tasked.

   b. *Law Enforcement.* Trackers may be called upon to assist the civil powers (MACP) in locating dangerous criminals who have gone into hiding in a rural environment. Tracker dogs and their handlers can also be used in follow-up operations.

   c. *Military.* It is within the military that tracking has the greatest application. The ability to identify sign, and pursue an enemy by the sign he has created has unlimited advantages, particularly in jungle counterinsurgency operations.

RESTRICTED

RESTRICTED

Purpose of Military Tracking

0104. In an environment where information on the enemy is limited and where the majority of movement is concealed the primary means of intelligence gathering will be patrolling. It should be the commander's intent to find, fix and strike at the enemy wherever he may be. The importance of tracking therefore cannot be underestimated as each and every sighting or contact with the enemy must be followed up in order to fulfil the commander's intent. The main purpose of military tracking therefore is to gain and pass onto commanders as much information as possible about the enemy by employing tracking techniques in order to mount subsequent military operations against them.

Tracker Platoon

0105. Operating in the jungle is manpower intensive and one of the factors which affects the manner in which jungle operations are conducted is the wide dispersal of force required to dominate the ground. However, the provision of sound intelligence is vital in mounting any operation and it is therefore recommended that each battalion should form a tracker platoon of at least three six men teams. The tracker platoon ORBAT would ideally be as follows:

Platoon HQ

Comd Officer or SNCO
Signaller
Driver
Storeman
Two Tracker Dogs and Handlers

Team 1	Team 2	Team 3
Commander	Commander	Commander
Signaller	Signaller	Signaller
Visual Tracker 1	Visual Tracker 1	Visual Tracker 1
Visual Tracker 2	Visual Tracker 2	Visual Tracker 2
Visual Tracker 3	Visual Tracker 3	Visual Tracker 3
Visual Tracker 4	Visual Tracker 4	Visual Tracker 4

0106. The options for establishing a tracker platoon are as follows:

a. In a jungle environment re-role elements of the Anti-tank Platoon as the tracker platoon. This option has obvious advantages as the armoured threat is greatly reduced therefore manpower could be provided without losing a capability elsewhere.

b. Train the Reconnaissance Platoon as a tracker platoon. This would require the Reconnaissance Platoon to fulfil two roles although it has a number of advantages as follows:

(1) The troops within the Reconnaissance Platoon generally have the high level of personal skills required by a tracker.

(2) The platoon is established and has all the equipment required by a tracker team (TT).

(3) On discovering an enemy objective the team would be experts in the conduct of Close Target Reconnaissance (CTR) and would not require additional training.

(4) The Platoon HQ is part of Battalion HQ which will speed up both the tasking of the platoon and the reporting of information.

c. Convert a section per company into a TT, and either keep them attached to that company or group them together as a platoon under Battalion HQ.

Tracker Team Tasks

0107. Whatever option is selected for forming a tracker platoon, the TTs will be employed in a similar manner. The tactics utilised by a TT whilst deployed are described fully at Chapter 4. The tasks of a TT generally fall into the following categories:

a. Information reporting.

b. Pursuit.

0108. ***Information Reporting.*** TTs may be employed purely to gather information and report back their findings. They may be inserted into an area on the basis of good intelligence, for example where enemy activity is either reported or expected, or possibly just to confirm that an area is free of enemy. In this task their role is to cast for sign in likely areas of enemy movement, and on locating sign follow it in order to gather as much information as possible. There are many reasons why a TT should be deployed on information reporting tasks of which the following are examples:

a. To patrol in support of an OP/reconnaissance screen.

b. To search for sign of the enemy in areas not covered by normal patrolling.

c. To follow up a sighting of, or intelligence on, the enemy.

d. To search areas where enemy activity is expected/reported.

RESTRICTED

e. To gain information on enemy tactics at an incident, i.e., to cast an area after an enemy ambush, or to search an old enemy camp location.

0109. ***Pursuit.*** This is when the TT is tasked as part of a follow-up force to pursue the enemy after a contact with a view of re-establishing contact. A tracker dog, which will speed up the pursuit, may be attached to the team. However the team will still be required to track visually. The team would be under the command of the fighting patrol commander and would normally lead the pursuit until contact is made or expected. A TT will be deployed for both planned operations and in reaction to situations as they occur. Examples of when a TT would normally be deployed in the pursuit are as follows:

a. *Pre-Planned Deployment.* A TT should be deployed as part of the task organized for immediate follow-up/exploitation purposes on the following operations:

(1) On a deliberate attack.

(2) On an ambush.

(3) During an advance to contact.

(4) On a cordon and search operation.

(5) In defence.

(6) On any fighting patrol task.

b. *Reactive Deployment.* A TT should always be available as part of a standby force to react to situations as they occur. The TT would normally deploy in order to pursue the enemy. The main reasons are likely to be:

(1) After a contact.

(2) After a sighting or on recently obtained intelligence.

0110. ***Employment.*** A TT due to the demanding nature of the task will inevitably contain soldiers with the highest standards of personal skills, operational discipline and fieldcraft, and must therefore be commanded by NCOs with high levels of initiative. Similar to the sections from the reconnaissance platoon, the TTs will be expected to deploy for long periods and to cover a great deal of distance with little or no direct support. A TT will be invaluable both as an intelligence gathering asset, and as a vital tool for seeking out the enemy. However, as TTs will invariably be in high demand they must be held as a battalion asset allowing teams to be attached to companies and commanded at that level for specific operations. This will ensure that the TTs are available for tasks of the highest priority.

Deployment

0111. As a general rule TTs should be deployed by the quickest means available. TTs should be proficient in deployment by foot, boat, helicopter and helicopter abseiling. A special harness is available for a tracker dog which enables it to be deployed by both abseiling and winching.

0112 – 0113. *Reserved.*

SECTION 2. — TYPES OF TRACKING

Introduction

0114. ***Definition.*** Tracking can be defined as the skill of being able to LOCATE, IDENTIFY, and PURSUE sign, and from INTELLIGENT INTERPRETATIONS and ASSUMPTIONS gain reasonably accurate information about the target concerned.

Types of Tracking

0115. The methods by which sign can be followed are:

    a. Visual tracking.

    b. Scent tracking.

0116. These two methods indicate that the sense of 'sight' and 'smell' are used. In addition, the senses of hearing and touch are used although the latter only to a small degree.

Visual Tracking

0117. Visual Tracking is the skill of being able to track a person or animal by the many signs and marks it has left. Examples are:

    a. Changes in the colour and the unnatural formation of vegetation due to disturbance.

    b. Bruises, breaks and cuts in vegetation.

    c. Water in areas that are normally dry.

    d. Lack of water or dew on vegetation.

    e. Mud or soil on grass or bushes.

f. Foot prints in muddy ground.

g. Sap from a bruised root or trunk of a tree.

h. Disturbance to animal, bird or insect life.

0118. Due to the fact that the visual tracker (VT) relies primarily on sight, he is unable to track at night except in very unusual circumstances, i.e., brilliant moonlight, artificial light in open country, or with the use of night observation devices.

Scent Tracking

0119. Scent tracking is normally performed by dogs, however, scent can also assist a well trained tracker in his pursuit of the target whether human or animal, providing the scent is strong and fresh. In addition the VT will use his sense of smell to warn him of any foreign scent in the area, for example:

a. Cooking smells.

b. Smoke.

c. Latrines.

d. Newly dug earth.

e. Perfume smells from toiletries.

0120 – 0121. *Reserved.*

SECTION 3. — QUALITIES OF A TRACKER

Introduction

0122. Generally trackers are most efficient in their country of origin; move them from their natural environment and their efficiency deteriorates until such time as they become accustomed to their new surroundings. However, there are certain fundamental factors that apply to tracking irrespective of the country, which are apparent to the trained tracker, whereas they remain hidden to the untrained. Nobody is born with the natural instincts of a tracker, however, contrary to the common belief a tracker is not a special type of person. He is a soldier who has attained these skills through dedication, practice and hard work. It is possible for anyone to become proficient in the skill of tracking provided they are determined and interested.

0123. ***The Tracker.*** The tracker is a highly trained soldier who is able to LOCATE, IDENTIFY and PURSUE sign and as a result form INTELLIGENT INTERPRETATIONS and ASSUMPTIONS about the target.

Qualities of a Tracker

0125. A competent tracker must possess certain qualities. These are:

    a. Honesty.

    b. Patience.

    c. Perseverance.

    d. Inquisitive mind.

    e. Acute observation.

    f. Determination (mental and physical).

    g. Above average endurance.

    h. Above average standard of fieldcraft.

    i. Knowledge of the enemy and his tactics.

    j. Knowledge about the local fauna and flora.

Summary

0126. With patience and perseverance it is possible for any soldier with the right attributes to be trained as a tracker. It is essential that a tracker has a sound understanding of the types of tracking, and the basic terminology used before he learns the elements of tracking.

RESTRICTED

RESTRICTED

Chapter 2

ELEMENTS OF TRACKING

SECTION 1. – GENERAL

Introduction

0201. Elements of Tracking (EOT) is a term used to describe the four main aspects of tracking, particularly when training trackers. The tracker should fully understand the elements of tracking, as these are the core skills or the fundamentals of tracking. The four main elements of tracking are:

    a. *EOT 1 – Sign* — which aims to define the characteristics, categories and classifications of sign.

    b. *EOT 2 – Factors Which Affect Tracking* — which aims to outline how sign is adversely affected by the elements.

Contents	
	Page
SECTION 1. — GENERAL	2–1
SECTION 2. — EOT 1 — SIGN	2–1
SECTION 3. — EOT 2 — FACTORS WHICH AFFECT TRACKING	2-4
SECTION 4. — EOT 3 — JUDGING THE AGE OF SIGN	2-10
SECTION 5. — EOT 4 — INFORMATION GAINED FROM TRACKING	2–13
Annexes:	
A. Observation and the Use of Other Senses	
B. Methods of Interpretation and Assumption	

    c. *EOT 3 – Judging the Age of Sign* — which illustrates procedures and considerations when determining the age of sign.

    d. *EOT 4 – Information Gained From Tracking* — which establishes exactly what information can be obtained through the application of tracking techniques.

0202 – 0203. *Reserved.*

SECTION 2. — EOT 1 — SIGN

Introduction

0204. Sign can be defined as 'any evidence of change from the natural state that is inflicted upon the environment by the passage of man, animal or machinery'. It is true to say that sign is visible to all however to the majority of us, 'the uninitiated', we do not always recognise the significance of what we are looking at. *See* Annex A.

Characteristics

0205. The ability to recognise sign is fundamental to being able to track, and the tracker who fully understands what sign he is looking for will be able to 'tune in' and

RESTRICTED

track in all environments, and in all weather conditions. All sign therefore may be identified by one or a combination of the following characteristics:

a. *Regularity.* Regularity is an effect caused by straight lines, arches or other geometrical shapes being pressed into the ground leaving marks not normally found in nature.

b. *Flattening.* Flattening is the general levelling or depression caused by pressure on an area, and is identified through a comparison with the immediate surroundings, e.g., bed space, boot print on grass, where someone has sat down, etc.

c. *Transfer.* Transfer or transference is a deposit carried forward over an area after the target has moved from one environment to another, e.g., mud, sand, grass, water, etc.

d. *Colour Change.* Colour change is the difference in colour or texture from the area that surrounds it, e.g., overturned leaves.

e. *Discardables.* Discardables are any materials 'cast off' (intentional or otherwise) by the target, e.g., rations, packaging, equipment.

f. *Disturbance.* Disturbance is any other change or rearrangement of the natural state of an area caused by the passage of the target, e.g., insect/animal life, dead leaves, bruised roots.

Categories

0206. Once the tracker has developed the ability to recognise the characteristics of sign he has acquired the primary skill required of a tracker. Whilst tracking all sign is divided into two categories with the dividing line taken at ankle height. The two categories are:

a. *Top Sign.* Is sign above the ankle to the height and width of the person and equipment he may be carrying. Examples of top sign are:

    (1) Broken twigs or leaves.

    (2) Scratches on trees.

    (3) Hand holds on trees.

    (4) Changes in colour and unnatural position of vegetation.

    (5) Cutting.

b. *Ground Sign.* Is sign left mainly by the feet or equipment placed on the ground. It refers to all sign below ankle height. Examples of ground sign are:

(1) Foot or boot marks.

(2) Broken twigs or leaves on the ground.

(3) Bruised or 'bleeding' roots.

(4) Disturbances of insect life on the ground.

(5) Disturbances of grass or ground vegetation.

(6) Mud, etc., deposited from boots.

(7) Disturbed leaves, stones and twigs on the ground.

(8) Discardables.

(9) Disturbed water.

Classification of Sign

0207. In any environment the tracker will be able to identify sign due to the fact that animals and indeed on occasions man may have been in the area prior to or since the target. It is essential therefore that the tracker is able to identify what sign belongs to his target and to discount 'foul sign'. Therefore when pursuing sign the tracker will designate all sign into one of two classifications as follows:

a. *Conclusive.* Conclusive sign is sign that indicates to a tracker the passage of his target through the area, e.g., a foot print, discardables or cutting; in other words sign that is definitely made by the target.

b. *Inconclusive.* This is sign that may or may not have been caused by the target (not definite).

0208. *Experience.* How a tracker determines conclusive and inconclusive sign will vary between trackers and will be dependent on the tracker's experience. A novice may require several characteristics of sign to be convinced he is still pursuing the same person/target. With experience and practice a tracker will require fewer or less characteristics to pursue his target and will treat what he once considered inconclusive sign confidently as conclusive sign.

Summary

0209. A good understanding of sign, and how to identify it, will assist the tracker during the initial learning stages of the course. EOT 1 is the most important part of visual tracking and it must be remembered that some people may not have the aptitude. With experience and practice locating and identifying sign will become easier.

0210 – 0211. *Reserved.*

RESTRICTED

SECTION 3. — EOT 2 — FACTORS WHICH AFFECT TRACKING

Introduction

0212. There are four main factors which can affect tracking and they are all very closely related. This can result in both advantages and disadvantages for the tracker. It is imperative that the tracker has a thorough understanding of each of the four factors, and anticipates how each one can affect the others. The four factors which affect tracking are:

    a. Sign (Spoor).

    b. Terrain.

    c. Climatic conditions.

    d. Time.

Sign (Spoor)

0213. Tracking the target's sign is the most important thing to consider but, in order to remain on the correct sign, a tracker must consider foul tracking by other human sign, or by animal sign as follows:

    a. *Animal Sign.* A tracker must consider the animal life in the area and know how to differentiate between animal and human sign. For example the majority of hoofed animals make a distinct chop-mark as they walk. The shape of the hoof acts like a knife-edge and the manner in which animals distribute their body weight onto the hoof causes it to cut into the ground. By contrast a man tends to heel and toe his movements when walking. As humans place their feet down the heel is the first part to touch the ground. As they are moving, their weight is being transferred from the heel onto the ball then onto the toe, which means the weight is being displaced evenly over the foot. For example, if a wild pig stood on a stick it would cause one clean break due to the cutting motion by which the pressure is applied, however a human would invariably cause two, one on either side of his foot due to the way he distributes his weight.

    b. *Other Human Sign.* If there is other human sign in the area, the tracker must determine through the sign pattern that he has identified which quarry to pursue.

Terrain

0214. The terrain in which the target is operating will have an impact upon the amount and type of sign produced. The characteristics are equally applicable how-

RESTRICTED

RESTRICTED

ever the tracker must have a comprehensive knowledge of what type of sign is likely to be prominent in different terrain and tune in accordingly. Tracking can be conducted in any environment. The following can be used as a guide:

    a. Grassland.

    b. Rocky Country.

    c. Primary Jungle (Rain Forest).

    d. Secondary Jungle (Scrub).

    e. Coastal and Estuarine Jungle (Rivers and Streams, Marshy and Swampy Ground).

    f. Sand.

0215. ***Grassland.*** If the grass is high, i.e., above three feet, tracks are relatively easy to follow due to the fact that the grass is knocked down and will stay down for some time, depending on the weather. If the grass is short it will spring back into its original position in a relatively short space of time. The following points will assist when tracking in grassland:

    a. Grass is normally trodden down and pointing in the direction the target is travelling.

    b. It presents a contrast in colour to the normal undergrowth when pressed down.

    c. If the grass has been wet with dew the night before, the dew will be rubbed off.

    d. Mud or soil from footwear may appear on some of the grass (transfer).

    e. If it is dry grass broken and crushed stems will be found. Footprints will normally be found in dry grass areas.

    f. If new vegetation is showing through it indicates that the track is an old one.

    g. In very short grass, i.e., up to twelve inches, a boot will damage the grass near the ground and invariably a footprint or impression will be found.

0216. ***Rocky Country.*** Tracking through rocky country is not as difficult as is imagined. This is due to the fact that rocks are both easily disturbed and are generally easily marked. The following are some points to consider in this type of country:

    a. Unless moving over large boulders, stones or rocks will either move aside or roll over. This will disturb the soil, leaving a distinct variation in colour and an

RESTRICTED

impression. If wet, the underside of the stone and rock will be darker in colour, and if dry, a much lighter colour.

    b. If moving over large stones the base of the boots tends to scratch the surface of the rocks.

    c. On sandstone, boot marks tend to show dark in colour, and on lava, the marks are whitish.

    d. If the stone is brittle it normally chips and crumbles when walked on and a light patch appears. The pieces of stone that have chipped off should also be seen nearby.

    e. Stones on the side of hills will move slightly or roll away when walked on, irrespective of whether the target is moving up or down hill.

    f. Stones on a loose or soft surface are normally pressed into the ground when walked upon, leaving either a ridge around the edge of the stone where it has forced the dirt out, or a hole where the stone has been pushed below the surface of the ground.

    g. Particles of stone sometimes catch in the sole of the footwear and are deposited further on and show up against a different background (transfer).

    h. Where moss is found growing on rocks a boot or hand will probably dislodge it.

0217. **Primary Jungle (Rain Forest).** Within primary jungle, trackers will find many mediums which readily show sign. These include for example the undergrowth, live and dead leaves, live and dead trees, streams with muddy or sand banks, and moss on the forest floor and rocks. Tracking in rain forest will be assisted by observation of the following details:

    a. Wet leaves on the forest floor when disturbed will show up as being much darker than those that are undisturbed.

    b. Dry leaves when undisturbed show a distinct bleached upper surface whereas in comparison the underside is dark brown. Therefore disturbed dry leaves will leave a significant dark colour change.

    c. Dead leaves often become very brittle and crack or break under the pressure of a person walking on them. The same can be said for small dry twigs.

    d. Where the undergrowth is thick, especially on the edges of the forest, the target may have to push through the vegetation. Bushes and branches with green leaves that have been pushed aside and twisted will expose the under side of the leaves which will be lighter in shade than the top of the leaf. To the trained eye this will be prominent. However, when looking for this type of sign the tracker must look through the vegetation and not at it.

RESTRICTED

e. Broken twigs will assist the tracker in assessing how long it is since the track was laid. Freshly broken twigs, green or dead, tend to portray a cream colour at the break and on fresh twigs fine hair like strands of wood will be present. The colour at the break will get darker and the fine strands dissipate with time, but if the tracker breaks the twig again he will get an indication by comparison of how long it is since the original break occurred. Only by experiment and experience will the tracker be able to determine the age of the break. Freshly broken green twigs usually retain the smell of sap for up to three to four hours.

f. Boot impressions will be left on fallen rotten trees if walked on.

g. Marks are generally left on logs which lay across a path or, if not on the logs, on the track on either side of the log.

h. Where roots are stepped upon they will often show bruising marks.

i. Broken cobwebs across the path may indicate passage of animals or humans along the path.

0218. *Secondary Jungle (Scrub).* This is the type of country where the primary growth has been cleared and the secondary growth has started. Usually, it is very thick and difficult to penetrate. To do so the individual may be forced to make his way through by either cutting or crawling along at ground level. When tracking in it, the main points to watch for are:

a. Broken branches and twigs.

b. Leaves knocked off branches.

c. Colour change, as in branches facing the direction in which the target has travelled.

d. Footprints on the ground which show up clearly as grass does not grow underneath.

e. Tunnels made very low to the ground.

f. Broken cobwebs.

g. Pieces of clothing or equipment caught on sharp edges of vines, bushes, etc.

0219. *Coastal & Estuarine Jungle.* The term 'Coastal' and 'Estuarine' jungle is taken to include rivers, streams, marshy and swampy ground. The most likely forms of sign to be located are as follows:

a. Footprints on the banks.

RESTRICTED

    b.   Footprints in shallow water or in mud.

    c.   Mud being stirred up, discolouring the water.

    d.   Rocks splashed with water in a slow moving/small stream.

    e.   Water on the ground at the point of exit.

    f.   Boots may have been taken off to wade the stream. Look for spots on banks where this was done and where they were put on again. Normally there will be sign where the person sat down to take off and put on his boots.

    g.   In mangrove swamps mud will be stirred up. Also branches of mangrove will be bent where people have held onto them to prevent themselves from tripping over roots.

    h.   Areas of colour change on river/stream beds indicating where the target has passed.

    i.   Top sign in the form of colour change where the target has pushed through tall reeds, etc.

0220.   **Sand.**   Sand is relatively easy to track on. The biggest problem to the tracker is that rain and wind can obliterate marks and impressions within a couple of minutes. The main points to be considered are as follows:

    a.   If the surface is reasonably hard, the footprint is very clear.

    b.   If the surface is soft the footprint will be quite deep and in the early morning and late afternoon the walls of the impression may cast a shadow.

Climatic Conditions

0221.   In tropical areas the weather can change rapidly between sun, rain and strong winds. Or, more frequently prolonged periods of either of these conditions can exist, such as in the monsoon and dry seasons. However each of these climatic conditions, either taken singularly or combined, have an adverse affect on sign and can therefore make the task of the tracker more difficult. The following climatic conditions should be considered:

    a.   *Direct Sunlight.*   The average temperature in tropical areas is between 30 – 40° C. In open areas such as grassland, or in exposed areas in the jungle, such as an LP or large bend in the river, direct sunlight will greatly affect sign. The heat will accelerate colour change to a far greater extent than in covered areas, for example an overturned leaf which is damp will quickly dry, or the darker side of an overturned dry leaf will become quickly bleached. However, in certain circumstances direct sunlight can be an aid, for example in disturbed areas it can produce a contained shadow.

RESTRICTED

b. *Strong Wind.* Strong winds are a frequent occurrence, especially prior to heavy rain, and although not always felt by troops operating within thick jungle, strong winds cause a great deal of disturbance to the vegetation. For example in exposed areas it will encourage disturbed vegetation to return to normal, i.e., blow it or the vegetation around it into different positions. In addition, strong wind may even conceal some ground sign by blowing vegetation on top of it.

c. *Heavy Rain.* The average annual rainfall in most of the tropical areas is around 500 cm (180") per year. Rain will occur frequently in primary jungle hence the name 'rainforest' however the majority of the rain will fall during monsoon seasons. Thick cloud will also reduce light therefore affecting visibility which will reduce the tracker's ability to track considerably. Heavy rain will wash out some sign very quickly, particularly ground sign, and is probably the tracker's greatest enemy.

Time

0222. To be able to assess the time between when the sign was first made and when it was located is the hardest part of a tracker's task. Full details on how to judge the age of sign are given at EOT 3 however experience and practice will help to overcome this difficulty. Obviously the longer the time span since the track was laid and the time of discovery, the more chance there is of the track becoming fouled or flawed by climatic conditions and as a result the harder the sign will be to locate and pursue.

Other Considerations

0223. In addition to the factors above, there are a number of additional considerations that will affect a tracker's ability. The two main considerations are as follows:

a. *Animal Life.* Within a jungle environment the amount and type of wildlife will both assist and hinder tracking operations. In many countries, birds, insects and mammals will warn other animals of impending danger from intruders. Monkeys are particularly proficient at shouting a warning when they see humans in the jungle. Insects on the other hand tend to go silent as a means of warning. A first hand knowledge of the habits and nature of the wildlife will undoubtedly assist a tracker in the application of his task. Large groups of animals between the tracker and the target can cause foul track so badly that pursuit becomes almost impossible. In this instance a tracker should consider applying the following drill:

(1) Look for a likely exit point.

(2) Isolate the track.

(3) Carry out a likely area search.

RESTRICTED

b. *Tracking in Built-up Areas.* Built-up areas including villages, roads, etc., cause additional problems for trackers. A track leading towards a village may indicate a RV with other forces, familiarity with the locals, or that the enemy target area is close. Tracking in these areas will invariably involve significant foul track and will slow down most trackers. However, the tracker may speed up his pursuit by carrying out a process of isolation procedures as follows:

    (1) Locate the entry and exit point.

    (2) Conduct a likely area search.

    (3) Look for the presence of the sign pattern being pursued.

Summary

0224. Each of the four main factors will affect the sign created by the enemy. It must be remembered that the type of sign will vary in different types of terrain. The tracker must take time to tune in and anticipate climatic conditions and time. The most important thing to remember is what can adversely affect tracking.

0225 – 0226. *Reserved.*

SECTION 4. — EOT 3 — JUDGING THE AGE OF SIGN

Introduction

0227. A tracker must be able to accurately judge the age of the sign encountered if he is to fulfil his role as an intelligence-gathering asset, regardless of whether he is information reporting or in pursuit. The ability to judge the age of sign accurately will only come with experience and continual practice however it is a vital skill which will assist the tracker in the following ways:

a. It will enable the tracker to determine the approximate time frame since the target passed through the area.

b. When tracking it is essential that the VT can determine if the target is in close proximity in order to prevent unnecessary casualties to the TT. Therefore judging the age of sign will assist the VT in deciding whether he is gaining on the target, or whether he is continually falling back.

c. By being able to deduce which sign is the same age as the target's sign will assist the tracker in overcoming the problems of foul track.

d. It will assist commanders in deciding whether or not the team should continue on track.

Considerations

0228. As stated in EOT 2 (Factors Which Affect Tracking) a VT must have an understanding of local flora and fauna and the relationship between these, as well as a knowledge of the weather and terrain in the area of operations. All these factors will determine the ageing process of sign found. When judging the age of sign the following factors must be considered:

    a. *The Type of Sign.* Sign will age to differing degrees dependent on whether the sign is hard or soft as follows:

        (1) *Hard Sign.* Examples of hard sign are marks in sun-baked soil, marks on stones, or marks in resilient mosses or tussock grass. Discardables such as plastic or metals are also considered hard sign. Hard sign takes longer to either deteriorate or return to normal therefore the tracker is likely to identify hard sign for a longer period than soft sign.

        (2) *Soft Sign.* Examples of soft sign are marks in soft soil, mud, or sand, and marks inflicted upon green leafy plants. Food discardables such as rice are also considered soft sign. Soft sign will deteriorate or return to normal more quickly than hard sign therefore the tracker knows that if he is identifying soft sign then he is closer to the enemy.

    b. *Exposure.* The degree of exposure to the elements will have differing effects on the sign. In areas exposed to direct sunlight and heavy winds sign will change rapidly (return to normal/age) which must be taken into account by the tracker. In exposed areas trackers may find that they are tracking on minimal sign and realise that when judging the age of sign any sign found is likely to be fresher than it appears.

    c. *Weather.* The condition of the sign will be directly connected with the weather conditions it has been subjected to. The tracker will know if the track was laid before or after heavy rain or showers.

Methods Employed

0229. Having considered the type of sign, its degree of exposure and the weather it has been subject to, the tracker could use one of the following methods to assist in judging the age of sign:

    a. *Comparison of Colour.* The VT should compare the colour of the sign in relation to colour of the surrounding area for colour change. Good areas to look for colour change are:

        (1) *Cracks in Bent Grass or Leaves.* An indication of the age of a track may be gained by the state of dryness of such cracks. When fresh they are green but after a few days they turn a brown colour. Therefore the older the sign the darker and dryer the crack is likely to be.

RESTRICTED

(2) *Breaking and Comparing Sticks and Twigs.* A fresh break will be a lighter colour and a study of ends will normally show fibres protruding. These fibres disappear over a period of time and their colour darkens therefore the fewer fibres seen and the darker the colour of the break the older the sign.

b. *Comparison of Impression.* The state of dryness of a track in mud or soft ground must be noted. If the track is very fresh, water will not have run back into the print or depression made by the target's foot. Later the water runs back and later still the mud that has been pushed up around the print and kicked forward begins to dry. By comparing your print beside that of the target's and then considering the colour change and the degree of regularity an assessment of the signs age can be made.

c. *Bracketing.* This method is based on knowing the weather conditions and the amount of wildlife movement typical in the area during the approximate period since the track was laid as follows:

(1) *Weather.* By considering the recent weather conditions it is possible to bracket the period when the sign was left for example:

(a) Time of Rain. A period of heavy rain will leave pockmarks on the sign, and will cause a loss of definition. Knowing the time of the last rainfall will provide a clue as to the age of the sign, for example if the last rain was 12 hours a go the sign would be over 12 hours old. Likewise, sign with little damage and plenty of definition will indicate that it was made after the last rainfall, i.e., less than 12 hours old.

(b) Overnight Dew. Where sign has traces of dew over it indicates that the sign was made the previous day.

(c) Strong Winds. High wind may result in leaves and foliage obscuring the sign. If this is the case and the tracker knows the time of the last heavy winds this can be a further aid to judging the age sign.

(2) *Game Tracks Superimposed.* Most wild animals lie up during the day and move at night. If human prints on a main game trail have animal tracks superimposed, and these tracks show that the animals have moved in both directions, then human prints are probably at least one night old or older. Likewise where human prints are superimposed on the animal sign then the track is likely to have been made that day.

d. *Local Knowledge of Flora/Fauna.* A detailed knowledge of local flora and fauna can assist the tracker in judging the age of sign as follows:

RESTRICTED

(1) *Vegetation.* The state and position of trodden vegetation must be noted. Various grasses have different degrees of resilience and prior knowledge of the flora's resilience to being crushed, pushed aside, broken, cut, etc., may assist in determining age. Only by studying flora present in areas of operations over a period of time can the tracker become adept with this method.

(2) *Wildlife.* By studying and noting the habits of wildlife, with regards to feeding, watering, movement, etc., the tracker can use this information to assist him in assessing age. As mentioned most wild animals feed and move during hours of darkness, but birds that feed on the forest floor, e.g., blackbirds, feed during the day so any sign disturbed by feeding or moving animals can assist in assessing age. Disturbed or crushed worm casts, spider webs, broken ant or termite trails should all be taken into consideration. Awareness of your natural surroundings will aid in judging age.

Summary

0230. Over a period of time all sign becomes subjected to natural processes which will eventually return it to a state where judging its age will become more of a guess. However, by considering the type of sign, the degree of exposure and the weather, and then by applying the methods to assist in judging age, it is possible for a VT to become adept at making accurate assessments of the sign's age. This will allow the VT to become an accurate intelligence-gathering asset.

0231 – 0232. *Reserved.*

SECTION 5. — EOT 4 — INFORMATION GAINED FROM TRACKING

Introduction

0233. In order for a tracker to make intelligent, accurate and logical assumptions about a target it is essential that he has a detailed knowledge of the enemy. Commanders at all levels must ensure that the tracker receives a full intelligence brief on the enemy prior to any patrol action. Quite simply the more the tracker knows the more he will find out. The target's mission will rarely be known at the beginning of a tracking task and will normally be determined by intelligence cells based upon both the information provided by the tracker and from previous information on the target. The information required prior to a patrol should come from the following sources:

a. Own intelligence cell.

b. Previous patrol reports.

RESTRICTED

    c.   Local population.

    d.   Police and other security forces.

0234. **Intelligence.** The tracker should conduct a detailed study of the enemy's background and in particular he should look at the following:

    a.   Enemy traits.

    b.   Enemy doctrine.

    c.   Enemy habitat.

0235. **Traits.** These are the attributes, of which many are inter-related, distinguishing a particular target from another. A study of the target's lifestyle will provide excellent information to the tracker as follows:

    a. *Attitudes.* This relates to the general moods or emotions of the enemy. Certain nationalities display particularly aggressive or emotional attitudes.

    b. *Beliefs.* This refers to the target's primary beliefs and principles. These beliefs are likely to influence the targets' behaviour. For example a fundamentalist enemy will often suffer more hardship than a less motivated enemy.

    c. *Habits.* This refers to the target's tendency or disposition to act in a particular way, for example always resting in the afternoon. These habits will become evident to a tracker over a period of time, and will normally assist the tracker in his pursuit.

    d. *Mentality.* This relates to how the target is likely to operate. Will he suffer hardship for a prolonged period, i.e., live on hard routine, or does he have certain expectations.

    e. *Physique.* This refers to the physical build of the enemy. Knowledge of the enemy's general physique (tall, small, muscular, slim, etc.) will assist the tracker to identify the target's sign in areas of foul track. It will also assist in determining the presence of foreigners amid a target group, i.e., the sign left by a European amid a group of Asians should be distinctive.

    f. *Diet.* This relates to food and drink that the target regularly consumes. An understanding of the target's diet will assist the tracker once again in identifying the targets sign, and may well assist him in making assumptions as to the targets routines and future intentions.

0236. **Doctrine.** Interlocked with his traits, consideration should be given to the target's doctrine, or teaching/thought pattern as follows:

    a. *Customs, Traditions and Religious Practices.* This relates to the target's usual or long-term habitual practices. Customs may differ from target to target, but a common thread amongst groups is likely to be evident. The target's traditions may well affect his daily movement and tactics, for example stopping to bathe daily or centrally feeding. A target's religious beliefs and teachings may even dictate halt times and rest periods.

b. *Military Practices.* A thorough understanding of the target's drills and procedures will assist the tracker on all tasks. A tracker should immediately recognise the type of enemy he is pursuing by the tactics employed and the numbers being tracked.

0237. **Habitat.** Finally a tracker should know where his target comes from. A study of the target's natural habitat will assist the tracker in his task by speeding up his ability to interpret information gained. In jungle regions the target's habitat would normally be one of the following:

a. *Remote.* A target from a remote region is most likely to possess excellent fieldcraft skills and be adept in living off the land however tradition and superstitions may well influence him. An enemy from a remote region may be difficult to track as he is more than likely to be able to track himself.

b. *Rural.* A target from a rural area would traditionally be a farmer, smallholder or peasant. He will in most cases have very good fieldcraft skills and will probably be able to travel light and live of the land. He will also be used to hardship which may result in him leaving minimal sign.

c. *Suburban.* A target from a suburban area will generally be far less aware of 'fieldcraft' and field skills than his rural counterpart. He will therefore usually be easier to track.

d. *Urban.* A target from an urban area will undoubtedly possess far fewer fieldcraft skills than a target from a remote or rural area. He is therefore likely to require more luxuries and will suffer less hardship than someone from another habitat.

Information Gained

0238. It is difficult for even small groups to move in the jungle without leaving sign noticeable to the trained eye. The tracker must be able to gain as much information as possible from any track. A VT, depending on the state of the track, can gain the following information: (*See* also Annex B)

    a. Direction of travel.

    b. Numbers being tracked.

    c. Age of the track.

d. Speed and load of people being tracked.

   e. Rations being used.

   f. Types of weapons carried.

   g. Tactics employed.

   h. Habits/Routine.

   i. Sex of target.

   j. General health.

Direction of Travel

0239. From a given length of track the VT can assess the general direction that the enemy is moving. Based on a good ground and map study the tracker/ intelligence cells can begin to identify the following:

   a. Possible enemy targets.

   b. Possible locations of enemy camps or resupply points.

   c. Likely enemy intentions.

0240. ***Direction Indicators.*** In order to ascertain the direction of travel a tracker must be able to identify those types of sign that can indicate the direction in which the target is moving. The following are used as direction indicators:

   a. Use of pointers.

   b. Transfer from one area to another.

   c. High grass and fern pushed down in the direction of movement.

   d. Heel marks or boot prints in favourable situations (indicator paces).

Numbers Being Tracked

0241. A knowledge of the enemy strength is essential and the tracker should try to gain this information as a matter of urgency. Determining the following factors may assist the tracker in estimating numbers:

   a. The number of sleeping areas identified in a harbour/LUP area.

   b. The number of resting places at a rest halt.

c. Different types of boot patterns (sign pattern).

d. By the use of a heel count in soft ground where boot prints are distinct using one of the following methods:

(1) Measure off a distance of 24 inches and count the number of heel marks noting that one man never places his feet down twice within 24 inches.

(2) Measure off one overstretched pace, then count the number of heel marks within that pace and divide by two. This will give you an estimate, i.e., five heel marks, two, possibly three persons.

(3) Measure off two normal paces, then count the number of heel marks within the two paces and divide by two. Again estimate; five heel marks, two, possibly three persons.

Age of Track

0242. Knowing the age of sign will enable the tracker to ascertain how far ahead of him his target is, and will enable both the tracker and intelligence cells to determine where the enemy may be when combined with knowledge of the enemy's speed of travel. Indicators to judge the age of sign include:

a. The state of surrounding vegetation.

b. Impressions in the mud (comparison).

c. Obliteration by rain (bracketing).

d. Cracks in bent grass or leaves (comparison).

e. Game tracks superimposed.

f. Leaves covering tracks (previous days climatic conditions).

Speed and Load of the Target

0243. Sign that indicates speed, and sign that indicates load, are intricately combined. Knowing the speed and load of the target is essential to the tracker/intelligence cells for numerous reasons, for example:

a. If the direction of travel is known, in order to gain on the enemy the track can be isolated by moving at speed off track before picking it back up and continuing on task.

RESTRICTED

b. If the overall plan is to destroy the enemy being tracked then knowing the speed and direction of travel will assist in determining where to site ambushes on his projected route.

c. Troops travelling light are likely to need resupply or early extraction. The tracker or intelligence cell must ask how, when and where will either take place. If either can be identified and patrols tasked early enough the likelihood of success against the enemy is significantly increased.

0244. *Speed and Load Indicators.* A tracker can estimate the speed and load of the enemy by utilising the following indicators:

a. The distance between rest/LUPs/meal halts will indicate how far the enemy can travel in one day or between meals.

b. In soft ground the distance between paces can indicate speed, i.e., large stride — fast moving, small stride — slow moving.

c. The depth of pace (in comparison to the tracker's) can indicate heavy loads being carried, which means they may be carrying some sort of pack.

d. Studying the amount of sign left behind will indicate speed, i.e., are they taking time to cover it up.

e. The route used and obstacles encountered will indicate speed and load. For example the enemy may go around an obstacle rather than over or through it, i.e., heavy loads, or the enemy may stick to ridgelines indicating a possible requirement to move quickly.

f. From the information gained regarding the speed of the enemy, the tracker must ask; are the enemy travelling light and moving fast, or are the enemy moving slowly and why?

Rations

0245. A tracker can gain the following information from discarded rations:

a. The nationality of the target if not known, i.e., certain foods may indicate origin. For example, continually finding rice might indicate the target is of Asian origin.

b. Who has supplied the rations, i.e., are they issued rations or locally supplied rations? If locally supplied where are the rations coming from?

c. Types of rations being used, i.e., heavy rations such as tinned food, or lightweight rations such as freeze dry products. By determining the type of rations used the tracker can ascertain the following:

RESTRICTED

(1) Lightweight rations may cause the enemy to LUP next to water or carry out frequent water re-supply drills. He will normally need to cook these rations.

(2) Heavy rations might indicate why the enemy are carrying heavy loads and perhaps why his speed is reduced. The enemy might also eat these rations cold and therefore leave less sign, i.e., fire marks, etc.

Types of Weapons

0246. If the weapons used are not known, identifying the weapons the enemy may be carrying is very difficult and will only be possible under favourable conditions. The following can assist the tracker:

    a. Information from a contact area such as:

        (1) Discarded ammunition cartridges.

        (2) Sign located during initial cast of incident area.

        (3) From friendly forces in the area of the contact.

    b. A sighting by a TT.

    c. Discarded ammo cartridges along the track/or in LUP areas.

    d. Flannelette located in rest/LUP areas, i.e., after cleaning weapons.

    e. Butt or bipod marks on the ground around rest/LUP/halt areas.

    f. Sight marks on trees where weapons may have been rested.

Tactics

0247. Identifying enemy tactics is a fundamental part of tracking especially when information reporting. A tracker must therefore determine the tactics employed by the enemy. Knowing the tactics employed will not only assist in determining how friendly forces operate against the enemy, but will assist the tracker in finding information. For example, knowing if the enemy posts sentries whilst lying up could affect when or how a follow-up force engages the enemy, or knowing that the enemy always travels with flank protection could determine how far from a track we site our ambushes. Likewise, the knowledge that the enemy snap-ambush before halting will assist the tracker when determining where to look for information, such as numbers of enemy, etc. Assessing the enemy's tactics will be done as the track picture is built up through intelligent interpretation of the targets:

    a. Tactics at halts.

RESTRICTED

RESTRICTED

    b. Tactics at obstacles.

    c. Tactics at LUP/Harbour areas.

    d. Deception tactics, if employed.

Habits/Routines

0248. A knowledge of the enemy's day to day patrol routine is vital to the tracker. After tracking the target for a sustained period idiosyncrasies may be detected which will both help to identify the patrol especially if an area of foul track is encountered and speed up the tracking process. The tracker should identify and recognize idiosyncrasies such as:

    a. The habitual siting of LUPs/harbours on or by certain topographical features such as streams. Knowing this will assist in identifying future enemy locations for either track isolation or follow-up purposes. In addition such a habit might also indicate the requirement for water for cooking or fishing for food.

    b. Negotiating obstacles or bypassing thick vegetation in a certain manner or direction. If this can be identified then ambushes can easily be sited in these locations.

    c. Carrying out halts or tasks at regular periods. This will assist the tracker in knowing where to look for incidents and will help him to anticipate enemy actions, i.e., meal halt followed by a comms halt, etc., thus speeding up his ability to interpret incidents and subsequently spend more time on track.

    d. Patrol members' idiosyncrasies may also be observed, e.g., smokers, twig snapping, etc., (sign pattern) which assist in identifying this patrol from other patrols.

Enemy Sex

0249. This is less likely to be identified than the other information gained but is detectable by the following means:

    a. Enemy forces having female soldiers will normally have separate sleeping areas and toilet areas, particularly those who have stricter moral codes, e.g., as was the case with the Vietcong and Malay Communist Terrorists.

    b. Females urinate squatting thus leaving a deeper depression in the earth between the feet than a standing male.

    c. Evidence may be found of the debris associated with female monthly periods.

RESTRICTED

RESTRICTED

General Health

0250. The target's general health can sometimes be detected by some of the following methods:

    a. Examination of toilet areas and faeces, e.g., diarrhoea or dysentery.

    b. Vomit or bloody sputum will indicate ill health.

    c. Blood stained bandages or other medical debris will indicate injury or ill health.

Summary

0251. The TT should always be on the look out for information. It is essential that the patrol receives a full intelligence brief prior to the action and they should report any information considered vital immediately. Each day a radio report will be sent including a summary of the information gained, and at the end of the patrol action the patrol should be fully debriefed and complete a full patrol report (*see* Chapter 5 for reports). Interpretation of the facts gained will start with the TT on the ground. However, intelligence cells must conduct their own close examination of the intelligence provided in order to ensure that vital information is not missed.

RESTRICTED

RESTRICTED

ANNEX A TO
CHAPTER 2

OBSERVATION AND THE USE OF OTHER SENSES

Introduction

1. The use of sight is the primary means by which a tracker gathers and assimilates information. An understanding of why things are seen will assist a VT in searching for and detecting enemy personnel, equipment and sign. In tracking terms the reasons why things are seen are as follows:

    a. *Shape.* Military equipment and the human body are familiar outlines to all soldiers. They can be recognised instantly particularly when they are in contrast with their surroundings. Distinctive shapes that are easily detected, unless concealed are helmets, webbing and personal weapons. When patrolling the tracker must, as well as being mindful of the track he is on, always move as a lead scout observing for sign of the enemy at all times.

    b. *Shadow.* To the tracker shadows provide advantages and disadvantages. Shadows commonly encountered are:

        (1) *Cast Shadows.* In sunlight/moonlight an object casts a shadow which may give away its presence. An object that is concealed in other shadows is harder to detect and does not cast a shadow of its own. As the sun/moon moves so do the shadows. Objects which were concealed by shadow may be revealed as that shadow moves. They may also be revealed by their own distinctive shadow which reappears. Trackers must therefore pay particular attention to the light around them and realise that in poor light they may miss some sign.

        (2) *Contained Shadows.* A contained shadow is contained within a space, for example, a room, a boot-print or under an individual shelter. It is normally darker than other shadows and can therefore attract attention. Whilst looking for sign disturbed vegetation may create a contained shadow especially in open areas.

    c. *Silhouette.* An object silhouetted against a contrasting background is conspicuous. An object may be silhouetted if it is against a background of another colour (a dark object against light back ground, light object against dark background). When searching incident sites the tracker should look for silhouettes by scanning and searching trees, banks, etc.

    d. *Surface.* If the colour and texture of the surface of an object contrasts with its surroundings it will be noticeable. Areas of flattening or disturbance will

RESTRICTED   2A-1

be identified because their surface is different to that of the ground around them.

e. *Spacing.* Natural objects are rarely, if ever, regularly spaced. Regular spacing draws attention. When examining incident sites the tracker should look carefully for areas that allow shelters to be erected for sign of enemy sleeping areas.

f. *Movement.* Sudden movement attracts the eyes. Whilst tracking in close proximity of a well disciplined enemy the tracker must remember that his movement may alert the enemy of his approach. His movement must therefore always be stealthy and he must observe for the enemy at all times.

Scanning and Searching

2. In order to detect enemy sign, a tracker must learn how to observe by scanning and searching as follows:

a. *Scanning.* Scanning is a general and systematic examination of an area, to detect any unusual or significant object or movement. Scanning applies throughout all stages of tracking, for example, when trying to assess the general direction of travel the tracker will scan the area to his front, or whilst tracking an incident site he will scan the area for initial sign. The main difference with scanning whilst tracking is the distance scanned. In close country it is greatly reduced, however, scanning over distance the ground must still be divided into foreground, middle distance and distance. Scan each area horizontally starting with the foreground at an incident site or the distance whilst conducting the track pursuit drill. To obtain maximum efficiency, move the eyes in short overlapping movements (moving the head will minimise eye fatigue however remember that movement must be kept to a minimum). The speed at which scanning is carried out will depend on the type of ground being observed, i.e., open or close country. When horizontal scanning is complete, scan along the line of the features which are angled away from the observation position.

b. *Searching.* Searching may take place at any stage during scanning. Any sign identified during the scan process requires an immediate search of the area. Whilst searching an area the tracker should be looking for sign, or any of the characteristics of sign, taking into account the reasons why things are seen.

The Use of Other Senses

3. In addition to observation, additional senses can be used to detect presence of the enemy as follows:

a. *Detection by Sound.* Within the jungle environment sound is usually the primary sense. Enemy camps and bodies of men on the move will normally be detected by sound well before they can be observed. Soldiers need to be trained to actively listen for sounds that are unnatural to the environment, particularly the noise of metal on metal, chopping of wood, movement and speech. Talking and similar sounds will be instantly recognisable to a trained soldier.

b. *Detection by Smell.* Although hearing and sight will provide most information, the sense of smell can also prove useful. The smell of cooking, campfires and tobacco are particularly noticeable, and there have been instances of camps being located by the smell of the sanitation pits. Odours will settle in low-lying areas and will hang for sometime in still air. Distinctive human odours are dependent on the diet. Soldiers whose primary food is meat will smell differently to those soldiers whose diet is made up of mainly rice and vegetables.

c. *Detection by Touch.* Although information will be gained primarily through the use of the other senses, soldiers need to be practised in the use of touch. A VT can gain valuable information from his sense of touch, i.e., cooking areas, sleeping areas. Detection exercises based on the use of touch not only lead to familiarity with shape and feel, but also encourage soldiers to use their deductive powers to gather information.

4. A VT must remember that his senses are his most valuable assets. It is important that trackers continually practise the skills utilised when locating the enemy, i.e., observation, detection by touch, detection by smell and detection by hearing. With practice the tracker will be able to recognise sign or enemy presence more quickly and will also obtain increased attention to detail of the environment, animal and insect life around him. At all times the tracker must be aware of the presence of the abnormal and the absence of the normal especially when close to the enemy or when operating against an enemy who uses booby traps.

RESTRICTED

ANNEX B TO
CHAPTER 2

METHODS OF INTERPRETATION AND ASSUMPTION

Introduction

1. As an intelligence-gathering asset, trackers must be capable of making logical and accurate assumptions about the enemy based on their interpretations of the sign and facts located on the track and at incident sites. This will be a continual process as the tracker builds a track picture. A tracker will not fulfil his mission if he is unable to carry out a logical process in order to reach these assumptions.

The Thought Process

2. The tracker will examine all information gained in the following manner:

    a. *Facts.* The tracker will identify facts. In tracking terms facts are defined as any area where an incident is known to have taken place or sign actually located along a length of track. Facts will normally be physical evidence for example a boot-print, an area of flattening or a discardable.

    b. *Interpretation of the Facts.* This is the logical thought process that the VT goes through at a particular incident or area of sign after studying all the facts and linking them together. In simple terms it is the VT's explanation and reasoning based on the facts as to who, how, why and when each incident or piece of sign was made. All of the above form part of a logical thought process. In common with an estimate a tracker would not be able to make an accurate interpretation without first establishing the facts.

    c. *Assumptions.* This is the tracker's general conclusion of the incident or area of sign. Having examined all the facts and having interpreted them the tracker is in a position to make an assumption as to what the target has done in the area, i.e., carried out a short term halt, or to anticipate the future intentions of the target.

    d. *Confirmation.* This is the verification of the tracker's interpretations and assumptions. Confirmation can be either:

        (1) Immediate confirmation by encountering the target or,

        (2) Delayed confirmation coming sometimes through locating further confirmatory facts, from intelligence reports, the local population, PWs or other means.

RESTRICTED

Summary

3. In order to gain the maximum information from any length of track, the tracker requires to interpret all the facts and make an assumption based upon them. Intelligence cells should always debrief tracking patrols thoroughly in order to ensure that the information reported is accurate and that nothing has been missed. The following format is advised.

    a. List all the facts gathered on the target.

    b. Make intelligent interpretations based on all the facts.

    c. Compile the assumptions into a patrol report.

    d. Conduct a thorough patrol debrief on completion.

RESTRICTED

Chapter 3

TRACKING TECHNIQUES AND PROCEDURES

SECTION 1. — GENERAL

"Remember that following up the enemy by tracking, is the most dangerous of all military tactics"

(Quote from Australian SAS Handbook)

Introduction

0301. As with normal patrolling operations drills and procedures are employed by TTs in order to ensure that all members know their role, and to negate the need for frequent orders and briefings. The use of drills therefore speeds up the tracking process. The drills included in this chapter are designed to ensure that tracking is conducted in a logical sequence thus ensuring that information is not missed. A TT must develop a thorough understanding of these drills if it is to be successful especially when operating against an enemy who is tracker aware.

Contents	Page
SECTION 1. — GENERAL	3–1
SECTION 2. — TRACK CASTING DRILL	3–1
SECTION 3. — TRACK PURSUIT DRILL	3–5
SECTION 4. — DUTIES OF A COVERMAN	3–8
SECTION 5. — INCIDENT TRACKING DRILL	3–10
SECTION 6. — TRACK ISOLATION DRILL	3–13
SECTION 7. — DECEPTION TACTICS	3–15

0302 – 0303. *Reserved.*

SECTION 2. — TRACK CASTING DRILL

Introduction

0304. In order to locate an enemy track, relocate a track after the sign has been lost, or after track isolation has been carried out, a tracker must carry out a sequence known as the track casting drill. As a general rule when tracking the tracker should never pass his last definite sign (LDS). If however the tracker can no longer see the sign he must employ the track casting drill utilising the techniques listed below in order to relocate the sign:

    a. Step 1 — Carry out an initial probe.

    b. Step 2 — Carry out an initial cast.

RESTRICTED

3-1

RESTRICTED

  c. Step 3 — Carry out an extended cast.

  d. Step 4 — Carry out a likely area search.

Initial Probe

0305. When the track has been lost and definite sign can not be located to the front the visual tracker (VT) must mark the last definite sign and carry out the following drill:

  a. Let the coverman (CM) know that he is going to carry out the initial probe. The CM will then move up to the LDS and provides cover for the VT.

  b. The VT then probes forward 3 – 5 metres in all likely openings until he locates sign. At the end of each probe he must mark the floor at the extent of his probe so he does not confuse his sign with that of the enemy (*see* Fig 1).

  c. Once sign is located the VT must confirm that the sign links back to the LDS then inform the CM and carry on.

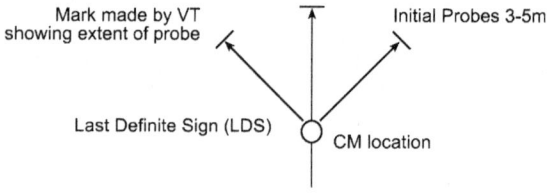

Fig 1. — Initial Probe

Initial Cast

0306. If the track is not located on the initial probe the VT must move back to the LDS and then carry out the Initial Cast drill (*see* Fig 2). The VT shows the patrol commander the position of the LDS then with the CM moves back from LDS along the side of track 10 to 15 metres to the rear of the LDS. From this point the VT and CM move off in a circle 10/15 metres radius around the LDS. Although the cast does not have to be a perfect circle it must be a complete loop of the LDS. The VT's aim here is to cut the enemy track. The direction of movement is optional, the VT usually being

RESTRICTED

influenced by the ground and general direction of the track. If the VT comes across the track he must confirm that it is the same track he has been following by checking the age and sign pattern. The VT then visually links up the sign with the LDS and then checks forward (away from the LDS) to ensure that it is not a dead end or deception. The VT then completes the circle in case of an enemy split up. During this drill the VT and CM should try and stay in visual contact with the rest of the patrol and the CM must remain alert, facing the direction of possible danger, listening and watching for sign or movement of the enemy.

0307. Before the VT and CM move off the commander moves the patrol up to the LDS and takes up a defensive position. They are briefed on both the direction out and expected direction of return of the VT and CM. On their return the information gained is passed onto the commander. The patrol now moves off and the VT confirms that he is on the right track. The time taken for this can range from 15 to 60 minutes depending on the enemy proximity, the ground and tracking conditions.

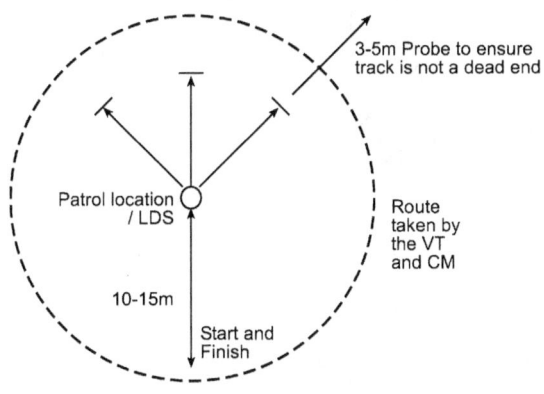

Fig 2. — Initial Cast

RESTRICTED

RESTRICTED

Extended Cast

0308. If the track is not located during the initial cast the patrol must conduct an extended cast (*see* Fig 3). Depending on the type of vegetation the patrol commander may allow the VT and CM to continue or elect to take the whole patrol on this cast. First of all the patrol or pair will move back another 10 – 15 metres from the start point used for the initial cast and then carry out a search by casting in a loop around the area of the LDS until the track is relocated. The total distance will be approximately 20 – 30 metres from LDS depending on the distance selected.

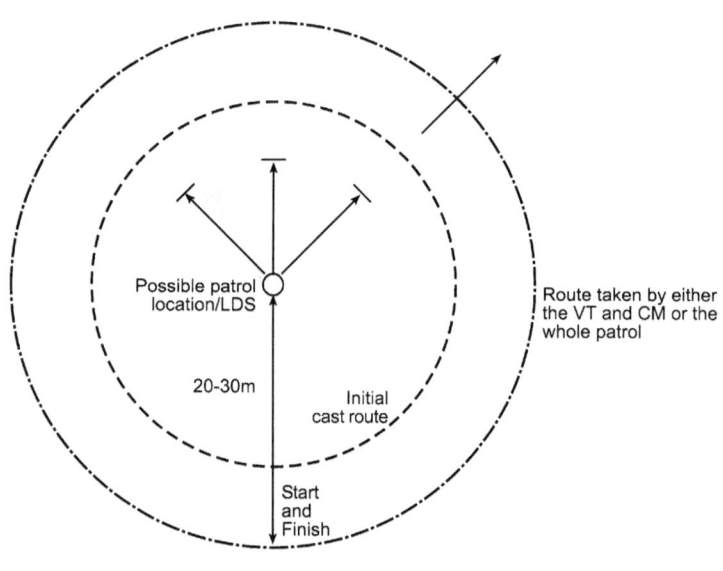

Fig 3. — Extended Cast

RESTRICTED

RESTRICTED

Likely Area Search

0309. If the extended cast is unsuccessful, the next step will be to conduct a likely area search. The tracker should make an assessment of the enemy's likely route and search the following places on and around it which are all conducive to leaving good sign:

    a. Streams and river banks.

    b. Areas of steep gradients.

    c. Likely harbour areas.

    d. River junctions and bridges.

    e. Road edges and tracks.

    f. Likely target areas.

    g. Villages or built up areas.

Summary

0310. The aim of the track casting drill is to assist the VT when locating or attempting to relocate the track. The VT must know where his last definite sign is at all times and carry out the drill properly. If the sign is not located during the initial probe, then it is very likely that the LDS may be further back along the track. Don't be tempted to probe too far as you will foul the area and reduce the effectiveness of the other drills.

0311 – 0312. *Reserved.*

SECTION 3. — TRACK PURSUIT DRILL

Introduction

0313. Once sign has been identified at the cast site the patrol can begin to pursue their target. From this moment they are considered 'on track' and should pursue the sign knowing that sooner or later they may encounter the enemy. It is essential that the VT remains on track at all times and must constantly observe the sign he is following, however, he must also be fully aware of the threat around him and therefore he must move with stealth and security at all times.

RESTRICTED

RESTRICTED

Track Pursuit Drill

0314. The tracker will track the enemy by employing the track pursuit drill. The drill is designed to ensure that the tracker focuses equally on the sign he is following, and on the enemy threat. There are five steps to the drill, each of which must be completed before moving onto the next. The tracker will always move as a lead scout and will conduct the drill as follows:

    a. *Step One — Assess General Direction.* The tracker should never move beyond the LDS. Once he reaches the LDS he should stop and observe around him looking for all possible openings in order to determine the general direction of the track (*see* Fig 4). To achieve this with some degree of accuracy, the tracker will mentally put himself in the enemy's place and say to himself *"Which way would I go?"* Sometimes the route will be obvious especially if moving along a severe ridge or through very thick jungle. If the TT has been on task for some time they will probably have already identified the target's general direction of movement which will assist them in this step.

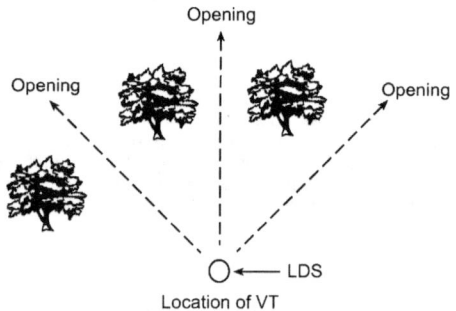

Fig 4. — Track Pursuit Drill – Step 1

    b. *Step Two — Eliminate Openings and Finalise Direction.* There will normally be a number of openings leading from the LDS (either natural or otherwise) other than the one attributable to the target. A tracker through training and experience, by a process of elimination, will be able to assess the most probable direction of enemy travel. In doing so the tracker must try to identify what is fresh or old sign and likewise what is human or animal sign. In areas where there is significant foul track the tracker will also attempt to eliminate openings by identifying his target's sign pattern.

RESTRICTED

RESTRICTED

c. *Step Three — Look For The Furthest Sign and Connect The Track Back to You.* Having identified the direction of travel the tracker must now begin to confirm the sign on track (*see* Fig 5). Initially he will identify the sign furthest away from him, i.e., the furthest sign he can see, then by searching the track back towards himself identify further enemy sign, thus connecting the track back to the LDS. It is not recommended to reverse this procedure as there is a tendency to make up sign where sign does not exist. It must be remembered that sign leads to sign therefore if he cannot link the sign back to him he should recheck the other openings.

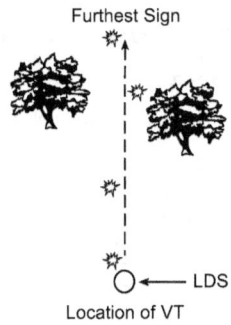

Fig 5. — Track Pursuit Drill – Step 3

d. *Step Four — Check The Areas Left and Right For Deception.* This step is designed to detect any possible deception tactics that the target may have employed and therefore a detailed understanding of enemy deception tactics is vital to the tracker. The CM will also check for deception especially whilst patrolling. The VT will check each side of the track by looking over his shoulder to the rear then scanning the ground to the flank along the entire length of the track to the furthest sign. The tracker then looks over his opposite shoulder and repeats the process. During this scan the tracker should also check for booby traps or booby trap markers. The tracker must remember not to confuse the enemy patrol naturally splitting to negotiate obstacles with an intention to conduct deception tactics. As a general rule the tracker will track the strongest sign.

RESTRICTED 3-7

e. *Step Five — Move Forward as a Lead Scout.* Having confirmed the track and checked for deception and booby traps the immediate area is then checked for presence of the enemy by scanning and searching to his front. The tracker now becomes a lead scout (*see* Chapter 4) and moves forward stealthily observing for the enemy until just short of the furthest sign. On reaching this position the tracker continues to scan the area and once satisfied that the area to his front is free from enemy he will repeat the five steps of the track pursuit drill.

Errors to Avoid

0315. In steps one to four the tracker remains stationary whilst he observes the ground. All unnecessary movement must be kept to a minimum in order to avoid compromise. Whilst on track there are five things that a VT must not do:

a. Don't look down when moving forward unless absolutely necessary, i.e., always observe for the enemy.

b. Don't make noise which will travel beyond visible distance.

c. Don't 'Bluff' yourself, only move on definite sign.

d. Don't grab vegetation as you will telegraph your position. On very steep inclines and declines if necessary grab the roots or the base of the tree.

e. Don't track on when exhausted.

Summary

0316. Conducting the track pursuit drill can be a time consuming process which is directly affected by the numbers being tracked and the professionalism of the enemy. As the tracker gains experience he will be able to perform the track pursuit drill with greater speed and efficiency however at all times he must ensure that he moves with stealth.

0317 – 0318. Reserved.

SECTION 4. — DUTIES OF A COVERMAN

Introduction

0319. The CM is normally a fully trained VT however if there is only one VT available within the TT the task of the CM may be carried out by a soldier who is used to acting as a lead scout. In order to provide maximum security to the patrol and to ensure that the tracking is of a high standard it is vital that a good working relationship exists between the VT and his CM.

Responsibilities

0320. The specific duties of a CM are as follows:

    a. *Provide Cover for the VT.* This is the primary role of the CM. Whilst carrying out the track pursuit drill whenever the VT is static observing for sign the CM is responsible for the provision of security to the front. The CM must therefore be positioned so that he is able to provide immediate cover before the VT carries out close examination of any sign. In simple terms the CM should take up a fire position covering the arc normally observed by the VT whenever the VT stops at the LDS to observe for fresh sign.

    b. *Provide Relief for the VT.* Tracking can be very taxing both physically and mentally. To minimise errors trackers should be relieved regularly. Generally the VT must be rested after a period of one to two hours of being on track. When handing over the VT should show the sign to the CM and inform him of what has been the key sign, any sign pattern and all details of the enemy. The CM must always take over on conclusive sign. If he is not happy with the sign at this stage the VT will continue on sign for a couple of further bounds until the CM is content to take over.

    c. *Check Left and Right for Signs of Deception.* Whilst moving the CM has the ability to observe for sign which the VT may have overlooked; this is because the VT is providing protection to the front. However, it is important that the CM does not try to do the VT's task but informs him of any sign that he locates. If the CM is not a qualified VT he must have a thorough knowledge of enemy deception tactics as this will assist him if deception is encountered along the track.

    d. *Note the Distance Travelled and Changes of Direction.* Although it is primarily the task of the patrol commander and 2IC to carry out navigational checks, the CM and VT should discuss any changes to the track direction that may indicate likely enemy intentions. It is essential that the patrol can state its position with accuracy as this will assist with both assessing enemy routes, intentions and with recording information.

    e. *Assist the VT During the Track Casting Drill.* During the initial probe the CM adopts a position close to the LDS observing the arc to the front thus covering the VT. If the VT fails to identify sign during the initial probe the VT and CM will conduct the initial cast, followed by the extended cast until the sign is relocated. During the casts the CM is responsible for providing cover to the VT as if he were carrying out the track pursuit drill. However, he must pay particular attention to flank protection.

RESTRICTED

Summary

0321. The security provided by the CM is vital to both the VT and the team. Acting as a CM is a mentally and physically demanding task which requires the individual to maintain a very high state of alertness for protracted periods. The CM must be prepared to take over from the VT at any time and in addition he must constantly observe for enemy deception and any other sign that the VT may have missed.

0322 – 0323.   *Reserved.*

SECTION 5. — INCIDENT TRACKING DRILL

Introduction

0324.   An enemy patrol will invariably carry out some form of routine drills and procedures. This will create what is known to a tracker as an incident site, i.e., an area of sign created by an enemy action that will provide significant quantities of information. In addition to a rest halt, LUP or harbour area, an incident site can be an area where enemy action of any sort has occurred, i.e., an ambush site. Reading the sign left by the enemy is the most important part of tracking. Each incident site will show some sign which will assist a tracker in reconstructing what has taken place. A thorough knowledge of enemy drills will assist the tracker in making intelligent interpretations at incident sites and ensure that all information is gained.

Incident Indication

0325.   Whilst pursuing a track a VT should always anticipate the likelihood of encountering an incident site. Incident sites may well have been disguised and could easily be missed if the VT does not consider the following:

a. *Track Picture.*   Having built up a track picture the tracker can begin to anticipate locations of possible incidents. Due to the distance travelled between LUPs, meal halts and any enemy patrol idiosyncrasies, i.e., the enemy stops at stream junctions.

b. *Direction Indicators.*   Noticing a direction indicator, i.e., the target breaks track, loops back, boot marks/pointers indicating that the target may have moved off his axis of advance but has now returned and continued in the original direction. These would indicate movement off of the track possibly indicating a routine drill, for example the enemy may always double fishhook before an overnight halt.

c. *Increased Sign.*   By identifying an increase in the amount of sign and disturbance in a particular area, i.e., flattening or the unnatural position of plants, leaves, needles, etc., a VT should be able to predict the presence of an incident site.

RESTRICTED

RESTRICTED

d. *Likely Areas.* The tracker should consider those areas where the enemy is most likely to stop, i.e., at streams and tracks for water resupply/obstacle crossings and at the top of steep hills for rest halts.

Sequence at the Incident Site

0326. The mission of the TT will influence the action taken at an incident site. A TT conducting a pursuit will want to move on quickly and will normally only confirm the entry and exit locations and continue on task. However in order to build a track picture a team that is information reporting will try to gain more information and will therefore track incidents as thoroughly as possible, although always mindful of not taking too long and possibly losing the enemy. By employing a logical sequence and a sense of urgency, relevant information can be gained without wasting precious time. The following sequence should be employed at the incident site:

a. *Move Back to a Start Point.* Once the incident is identified/suspected, the VT should inform his commander and then move back and adopt a position of all round defence short of the incident site, similar to when conducting an initial cast. If on first impression the incident site appears large and it is anticipated that the search may be prolonged sentries with claymores should be established.

b. *Plan the Search.* Once secure the VT will confirm his plan with the commander which must start by looping the site to determine the size, layout and entry/exit points of the incident. This may include:

(1) Dividing the area into segments to assist in searching.

(2) The deployment of a second pair of trackers.

c. *Conduct the Search.* The VT and CM conduct a systematic search compiling a list of all facts and draw sketches of the area including the north pointer, entry and exit points, the overall size and the position of all facts located (*see* Chapter 5 for layout). The search is conducted in the following manner:

(1) *Cast the Area.* The VT and CM conduct a full cast, looping around the area in order to determine the full extent of the incident. As with an initial cast (*see* Section 3) the VT will loop either left or right around the area, however he must endeavour to observe the sign in the incident site at all times so as to determine the full extent of the area without straying to far away. It is critical during this cast that the VT identifies any exit points or routes used for water, latrines, etc, and marks his own sign accordingly.

(2) *Divide the Area Into Segments.* Once the full extent of the area has been determined the VT should position himself at the entry point and then split the area into separate search areas. From the entry point divide the

RESTRICTED

area using the clock method. Entry point 6 o'clock moving either left or right, 9 o'clock through 12 o'clock or 12 o'clock through 3 o'clock then back to 6 o'clock.

(3) *Search Each Area.* Whilst utilising a sketch the VT should record all the facts identified and note his interpretations against each fact or group of facts. He then debriefs the team commander and any other trackers, giving an overall assumption of the incident.

(4) *Record/Report the Information.* The detailed recording of the incident will be conducted at the end of the patrol day in the tracking report (*see* Chapter 5) in the patrol LUP or base location. Any critical information however must be sent immediately.

Considerations

0327. Whilst tracking an incident the following must be considered by all:

a. *Time.* The VT and team commander should ensure that sufficient time is spent on a thorough and systematic search of the incident area. The time used constructively at this stage may prove invaluable in the event of the target sign being lost due to the weather or foul track. It will also help to identify a target group if track isolation is considered or conducted in the future.

b. *Individual Awareness.* All individuals must remain fully aware of enemy activity whilst at an incident. Booby traps and stay behind parties along with indirect fire are just some of the threats that can face a TT whilst at an incident.

Summary

0328. All the information listed at EOT 4 (*see* Chapter 2) can be gained from an incident site, i.e., direction, numbers, age, speed and load, rations, weapons, tactics, habits and routines, sex and general health. The time spent on the search at an incident area will vary according to the type and size of the incident and the information already gained on the enemy. The tracker will take longer at incidents identified early in the pursuit in order to build up a track picture. However later on when he has identified a number of similar incidents, i.e., several short term halts, and therefore developed his track picture he will be able to continue on track more quickly at this type of incident as he will already have most of the information to be gained.

0329 – 0330. *Reserved.*

SECTION 6. — TRACK ISOLATION DRILL

Definition

0331. Track isolation is where the tracker, having anticipated the intended route of the target, is able to 'leap frog' ahead and through the sign pattern of his target and identify the same track.

Application

0332. Track isolation is used on a number of occasions although predominantly when foul track is encountered or when the tracking team is attempting to make up ground on the enemy. Track isolation should be applied as follows:

    a. When foul track is encountered.

    b. When the position/route of the enemy can be predicted (for example through an assessment of his route/direction of travel, mission or intelligence received).

    c. In order to avoid contact or compromise with other enemy or civilians.

    d. When tasked/ordered to gain ground on the enemy.

    e. When the route of travel is obvious, i.e., on a steep sided ridge.

    f. In order to avoid difficult terrain.

    g. In order to overcome deception.

    h. In order to move on to fresh sign.

Predicting the Enemy Location

0333. On the majority of occasions the length of track isolated will be short and the tracker will simply move at a greater speed in the direction of travel and then regain the sign and continue on task. Greater distances however can be isolated if the tracker is able to predict/anticipate the likely position of the enemy. This is not a simple task although in some areas where movement corridors are restricted it may be simpler. The ability to predict where the enemy is going will come with experience and a thorough knowledge of the enemy's method of operations and an understanding of his likely mission or task. In predicting the likely objective or route taken by the enemy the TT will normally be assisted or even directed on some occasions by the Battalion Intelligence Cell. The following factors should be considered when attempting to assess the enemy's likely route/position and a combination of the information provided from considering each factor will normally point to the enemy location:

RESTRICTED

a. *Speed and Direction of Travel.* When the tracker has identified the direction of enemy travel he should ask "what distance is the enemy travelling per day" and, having identified the age of sign he is on, predict where the enemy is as follows:

    (1) Age of sign — e.g., 48 hours old (two days).

    (2) Distance travelled per day — e.g., 3 km.

    (3) Direction of travel — e.g., North.

    (4) Likely location of the enemy is therefore 6 km North of the current location.

b. *Terrain.* The tracker should look at the terrain and determine which would be the most suitable for the enemy he is following, i.e., knowing his method of operations think which way would he go.

c. *Track Picture.* The track picture identified will assist the tracker in determining where the enemy is: for example, the type of location where he lies up, i.e., next to a stream, or the navigational features he uses, i.e., stream junctions or main features such as ridgelines.

d. *Enemy Tactics.* The strength and professionalism of the enemy patrol should also be considered as this will assist in determining where the enemy may be. Tactics will vary depending on their mission and strength. Reconnaissance patrols will normally avoid good movement routes, such as ridgelines, whereas fighting patrols will probably use them.

e. *Intelligence.* Intelligence of your own and enemy force locations when combined with information gained on track will assist in indicating the route and objective. An enemy carrying heavy stores may well be travelling to a resupply point, whereas a possible reconnaissance patrol may be heading towards a friendly force location.

0334. Having considered the above the TT will be able to conduct track isolation and pick up the track at a later stage. When the enemy target is identified, or mission assumed, all information should be reported immediately. As previously mentioned the identification of the enemy target, route, mission is also the task of the Intelligence Cell.

Method of Relocating the Track

0335. The TT will decide where to relocate the track and will move to that area. On arrival the VT and CM will cast for sign as per the track casting drill. It is imperative however that the target is identified by his sign pattern as quickly as possible. Confirmation through the enemy sign pattern is the only way of being certain that the same target is now being pursued. For example, this may be clarification of the strength, speed and load of the target, or any idiosyncrasies identified.

Summary

0336. Once a comprehensive track picture has been obtained the TT employing the track isolation drill will have little difficulty relocating the enemy. Track isolation is a useful tool for gaining on the enemy, in particular when pursuit tracking.

0337 – 0338. *Reserved.*

SECTION 7. — DECEPTION TACTICS

Introduction

0339. VTs must be aware that a determined and professional enemy will undoubtedly employ deception tactics as an anti-tracking measure, especially if they are aware that trackers are likely to be employed against them. Deception tactics are designed primarily to slow down the tracker and prevent him from gaining ground and information. There are many ways that a determined enemy can confuse, delay and possibly lose a pursuing TT, therefore all VTs must have a thorough understanding of the main deception tactics used. It is vital that each VT is able to recognise evidence of deception immediately in order to remain on track and not lose ground on the enemy.

Deception Tactics

0340. The enemy may employ any or a combination of the following deception tactics:

    a. Walking backwards.

    b. Conversion of sign.

    c. Brushing the track.

    d. Stone hopping.

    e. Fade out.

RESTRICTED

    f.   Crossing or walking in a stream.

    g.   Splitting up.

    h.   Walking along a log.

    i.   Doubling back or backtracking.

0341. **Walking Backwards.** Walking backwards will normally be employed where the ground is soft and footprints can be easily identified. When the target walks backward he is attempting to confuse pursuing trackers as to his direction of movement. The tracker must note that when walking backwards the length of his pace is shortened. The toe and the ball of the foot will be more pronounced than the heel, the opposite of walking forwards. Loose dirt, dust, sand, and leaves will all be dragged in the direction of movement and in certain environments there should still be a significant number of pointers indicating the genuine direction of movement.

0342. **Conversion of Sign.** Some enemy may attempt to disguise their sign to look like animal tracks. Unless the enemy is extremely proficient in this method it will only serve to sign post his direction of travel. It is known that previously enemies have tried to conceal their sign by making animal tracks over them, namely pigs, by the use of shaped sticks. Conversion of sign of this nature is not to be confused with normal animal movement. Alternatively the enemy may choose to travel along a well-used route in order to conceal his sign amongst that imposed by other forces. The tracker should attempt to either isolate the track and relocate the target through his sign pattern, or continue on task searching for his exit point.

0343. **Brushing the Track.** The enemy will employ this tactic in areas conducive to leaving good sign, for example, on a muddy bank after a river crossing, or simply to cover up leaves along his route or at a break-off point. To an experienced tracker brushing the track is rarely successful and often tends to signpost the enemy's intentions. The tracker must remember that sign leads to sign and at some point when brushing the track has been employed the sign he has been following will disappear. A track that has been brushed will look different to its surroundings and once the tracker has tuned in will identify it by either disturbance or colour change.

0344. **Stone Hopping.** In this method the enemy attempts to move from large stone to stone over a piece of ground, especially besides or across a stream or river. If the enemy is successful in reaching each stone this tactic can cause a tracker problems. With perseverance however, a tracker should be able to identify enough sign to continue. For example, after a stone has lain for some time, dirt and sand will build up around its base, once stepped on this wall will crumble and a shadow will appear around the base, (a gap between the stone and the dirt). Scratches or marks, fine particles of dust, sand or dirt may also be left on the surface of the stone.

0345. **Fade Out.** When employing fade out as a deception tactic the enemy patrol will attempt to fade out his sign over a period of time. This is achieved by dropping off

RESTRICTED

one man from the patrol at a time, i.e., at 20 – 30 metre intervals. Individuals will then continue to step off the line of march before joining up again some distance away, or will all backtrack some distance to a single jump off point. Through the correct application of the track pursuit drill (step four, check left or right for signs of deception), an alert VT should notice the amount of sign he is following decrease. The CM who checks the sides of the track may well notice the 'jump off points'.

0346. **Crossing or Walking in a Stream.** The enemy may attempt to cross streams and rivers or may move along a stream or riverbed for some distance in order to minimise their sign. In some streams especially where the water is fast flowing or murky this can result in the enemy leaving little or no immediately visible sign. In clear, fast water the lack of visibility may be overcome by the use of a glass/plastic bag enabling the VT to examine the streambed for sign. This process will normally be very slow and a more likely alternative will usually be to concentrate on locating the exit point from the water where significant quantities of sign are often left. In clear water visual tracking is more feasible. Sign such as colour change, i.e., broken stones or marks on the riverbed, disturbance of the riverbed, and often pointers in reeds and rushes will enable the VT to remain on track. The tracker's greatest hindrance whilst tracking in water is heavy rain resulting in flash floods or fast currents which may well wash sign away. If this is the case the VT will have to extend his search in order to locate the enemy's exit point.

0347. **Splitting Up.** By splitting a patrol into two or more groups an enemy will leave significantly less sign. Often the patrol will only split for a short distance, but may sometimes split and meet up at an RV point later. When splitting up is encountered the tracker should follow the sign that is the most prominent.

0348. **Walking Along a Log.** Where a log or large piece of deadfall lies across or adjacent to the enemy line of march they may attempt to move along it, trying not to leave any sign as they do so. This will rarely cause a problem to a tracker who should be able to locate bruising, disturbance or transfer on the log, or as an alternative identify the jump-off point.

0349. **Doubling Back or Backtracking.** The enemy may often attempt to throw a tracker by laying a track in their general direction of travel, before moving back along it to a jump-off point some distance back. This type of deception is often combined with an attempt to conceal the jump-off point by either brushing the track, walking along a log, etc. It is also feasible that not all of the target may have laid the deception leg. The enemy's intent will be for the VT to continue on the false sign and eventually come to a dead end without realising that the enemy has jumped off the track and will therefore have to go back and search for the jump-off point. A worse case is if the enemy track leads to an area not conducive to leaving good sign, i.e., rocks or a murky stream, the tracker may be off track for a considerable amount of time before he notices the deception and finds the jump-off point. There are certain signs that the tracker should look for which illustrate backtracking, e.g., pointers facing towards the tracker.

Summary

0350. It is essential that a tracker has a complete knowledge of deception tactics and is able to recognise their use early in order to avoid losing ground on the enemy. A good enemy may well combine several deception tactics in one area. Evidence of deception should indicate to a tracker that the enemy is aware of TTs in the area, and this will obviously have an effect on his movement and tactics as the use of booby traps and stay behind parties should also be anticipated. Fundamental to his success in defeating deception tactics will be the correct application of the track pursuit drill and the assistance offered by the CM observing for deception as he moves forward. In addition trackers must continually train in anti-deception measures, i.e., tracking in streams, in rocky areas and on logs in order to become proficient at locating sign in these areas.

RESTRICTED

Chapter 4

TACTICAL TRACKING

SECTION 1. — GENERAL

Introduction

0401. The drills and procedures listed in this chapter should be applied in conjunction with the basic principles of patrolling and the routine drills contained in Infantry Tactical Doctrine, Volume 1, Pamphlets No. 3 and No. 5. This chapter is designed to illustrate both how a TT conducts its routine patrolling and how it conducts a pursuit. When employed on an information reporting task the TT will work purely under the command of the team commander. However, it is when the team is employed in the pursuit attached to a follow-up force that a greater understanding by commanders of tracking techniques is required.

Contents	Page
SECTION 1. — GENERAL	4–1
SECTION 2. — TRACKER TEAM COMPOSITION	4–1
SECTION 3. — DUTIES OF A LEAD SCOUT	4–5
SECTION 4. — TRACKER TEAM ROUTINE DRILLS	4–7
SECTION 5. — THE PURSUIT	4-12
Annex:	
A. Tracker Team Kit and Equipment	

0402 – 0403. *Reserved.*

SECTION 2. — TRACKER TEAM COMPOSITION

Introduction

0404. In general terms any TT is capable of conducting either information reporting or pursuit tasks. Usually a team tasked to conduct an information reporting patrol will consist of VTs. The addition of a tracker dog will assist the VTs especially in urban areas. Whilst for a pursuit a tracker dog is attached to the team. Anything other than this will reduce the TT's efficiency. A TT consists of three elements: the command, the tracker and the protection element.

0405. The TT will be a valuable information gathering and search asset. All team members must be physically robust and extensively trained as VTs. In order to conduct their task the team will have to be close knit and well motivated as they will be required to travel light yet operate in the field for indefinite periods. In addition the team must be proficient at insertion by any means whether by foot, boat, air or vehicle. Annex A is a suggested scaling of TT equipment.

RESTRICTED

RESTRICTED

Organization

0406. The ideal TT is organized and responsibilities allocated as follows:

    a. *The Command Element.* The command element consists of the team commander and a signaller. It is not necessary for them to be tracker trained.

    b. *The Tracking Element.* The tracking element consists of two VTs. The first VT will be employed as the lead scout (*see* Section 3) and will carry out the track pursuit drill (*see* Chapter 3). The other VT will be employed as the CM (*see* Chapter 3). On occasions the tracking element will include one or two tracker dogs and handlers, with one or both VTs acting as CM for the dog and handler.

    c. *The Protection Element.* The protection element consists of a further two VTs who will relieve the tracking element as and when required. If fully trained VTs are not available then riflemen who are used to operating as lead scouts may be used.

Capabilities and Limitations of a Tracker Team

0407. **Capabilities.** In general terms a TT has the following capabilities.

    a. *Deployment.* The team inclusive of a dog can be deployed by any means including helicopter abseil.

    b. *Duration.* The team can ideally carry food and equipment to operate for up to five to six days. Any increase in duration without resupply will add weight and slow down the follow-up. Providing the team can be re-supplied they can continue to operate indefinitely.

    c. *Age of Track.* When tasking a VT sign up to 72 hours old should be used as a planning figure. This time will vary depending upon conditions. For a tracker dog 24 hours is a good planning figure although up to 72 hours is not beyond a dog's capabilities in favourable conditions.

    d. *Communications.* The team must be capable of communicating on both VHF and HF nets.

    e. *Intelligence.* The team is able to locate, identify and pursue sign and from intelligent interpretation of the facts pass on information to commanders.

    f. *Reporting.* The team is able to report the information it has gained by either radio, patrol report, hot debrief or presentation (*see* Chapter 5) depending on the requirements of the commander.

    g. *Pursuit.* The team is able to pursue sign following either a contact or siting in order to gain or maintain contact as required by the commander.

0408. **Limitations.** In general terms a TT has the following limitations:

    a. *Night.* The team is unable to track at night through close country in a tactical manner.

    b. *Weather.* Rain and weather extremes will degrade the sign/scent left by the enemy and will therefore slow trackers down.

    c. *Contact.* The TT is vulnerable to contact when tracking an enemy who is tracker aware.

    d. *Time.* A track that is more than 24 hours old will slow the pursuit.

    e. *Foul Track.* In areas used frequently for patrol or civilian movement enemy sign will be difficult to identify due to tracks being overused.

Visual Trackers and Tracker Dogs

0409. When a tracker dog is included within a TT it will be the decision of the team commander as to whether he tracks using the dog or tracks using a VT. When making his decision he will consider the following:

    a. *The Tracker Dog.* In accordance with the information contained in Infantry Pamphlet No. 5 a tracker dog offers the following advantages and disadvantages:

        (1) *Advantages.* These are:

            (a) It is normally faster than a VT.

            (b) It can track over terrain which for practical purposes does not hold visual sign, i.e., a road.

            (c) It can track at night if night movement is possible.

            (d) The tracker dog will indicate when nearing the enemy. The distance will vary depending on the environment.

            (e) A tracker dog is less susceptible than a VT to foul track because it tracks by scent. Once it has built up its own track picture the scent of the enemy cannot be lost accept for in bad weather or after a prolonged period.

        (2) *Disadvantages.* These are:

            (a) When required to work under difficult conditions its tracking ability may deteriorate.

RESTRICTED

(b) It cannot give verbal information regarding the track that it is following. It is only as good as the handler and the handler may provide limited information on the track.

(c) The dog may have an off day.

b. *The Visual Tracker.* Similarly a VT has advantages and disadvantages as follows:

(1) *Advantages.* These are:

(a) He is able to give verbal account of the information. This may include numbers, quality of the enemy, weapons and other equipment carried.

(b) He is able to provide mutual support for the tracker dog and handler, also assisting in relocating sign if it is lost.

(c) Even when not tracking his superior powers of observation are invaluable.

(2) *Disadvantages.* These are:

(a) A VT cannot track at night without some form of artificial lighting and then only with limited success.

(b) He is normally slower than a tracker dog in pursuit.

(c) Any early warning given will normally be when the enemy is very close.

0410. **The Infantry Patrol Dog.** The Infantry Patrol dog is not a tracker dog but is trained to give early warning of the presence of human beings in a patrolled area. This class of dog, although not part of the TT, can be of assistance to them under certain conditions. *See* Pamphlet No. 5, Infantry Company Group Jungle Tactics.

0411 – 0412. *Reserved.*

SECTION 3. — DUTIES OF A LEAD SCOUT

Introduction

0413. One of the most physically and mentally demanding tasks in the jungle environment is that of the lead scout because as an individual, both his personal patrol skills and his powers of observation have to be of the highest calibre. Whilst tracking an enemy the likelihood of contact is greatly increased for at any time the TT may encounter the enemy they have been tracking. Due to the restricted visibility in close country, encounters with the enemy can happen at very close quarters with little or no warning. The action of the lead scout therefore at this critical point of contact can be literally a matter of life or death. Acting as a lead scout goes hand-in-hand with being a VT as he will not only track the enemy but will have the added burden of acting as the lead scout for the team. The VT will be supported by his CM and the importance of their having a good working relationship cannot be emphasized enough. They must work in unison with the CM protecting the VT at the appropriate times so that the VT can conduct the track pursuit drill if the team is to maintain its security. The duties of the CM and the track pursuit drill are covered in detail in Chapter 3; the detailed responsibilities of the lead scout are covered in this section. There are many occasions when a VT will act as a lead scout but will not necessarily be tracking, for example when patrolling to a cast site, or when patrolling to a LUP.

Considerations

0414. *Equipment.* Due to the taxing nature of being a lead scout and the requirement for him to have heightened senses, his equipment/load needs to be kept to a minimum. He should not be carrying any of the team equipment such as the radio, medical kit or any support weapons.

0415. *Position and Spacing of the Lead Scout.* The scout is the eyes and ears of the team and 90% of the time a contact is going to come from the front, unless a well-laid ambush is encountered. Therefore the scout's position and spacing is critical; it goes without saying that the scout is going to be the first man in the team. However, his spacing is the responsibility of the second man. Normally this spacing will be slightly greater than that of the remainder of the team enabling the scout to use his senses fully. However, when tracking at times the CM must move further forward in order to provide security for the VT. This difference is only marginal and the CM will constantly be adjusting the distance depending on the visibility.

0416. *Rotation of the Lead Scout.* The lead scout can very rapidly become both mentally and physically exhausted due to the nature of the task. Commanders need to recognize this fact and consequently rotate the lead scout on a regular basis.

Duties

0417. **Arc of Observation.** The primary arc of the lead scout whilst moving is roughly 90° either side of his line of advance. He should not be concerned about what is happening to his rear; that is the task of the remainder of the team although he is still required to look back occasionally for any signals or messages being passed to him.

0418. **Searching the Ground and Other Senses.** In close country hearing is the primary sense. However, sight will more than likely be the scout's first indication of a well-trained enemy. He should not ignore his other senses; hearing and smell are also going to assist him, therefore regular listening halts should be used throughout a patrol. It is vitally important that the scout has the ability to observe correctly in close country. Moving the head and scanning to his front too fast will only lead to him seeing a green blur. He should slow his actions and move his eyes as well as his head. He should endeavour to observe through the vegetation rather than looking at the vegetation. He is searching for the slightest movement, shadow or any number of signs that are going to alert him to the presence of the enemy.

0419. **Hand Signals.** The use of hand signals negates the necessity for the commander to constantly have to stop the team to talk to the scout. It is therefore important that the scout has a thorough knowledge of all the hand signals in use and when using them does so in a slow deliberate manner below shoulder height, so as not to alert an enemy to his position by rapid, jerky movements.

0420. **Navigation.** As the scout is concentrating all his energies on the security of the team his ability to keep an accurate record of the team's location is somewhat harder. It is important that the team commander up-dates him with the team's exact location as regularly as possible. He also needs to be informed of all ERVs on route, should the team have a contact or get split for other reasons.

0421. **Route Selection and Ground Awareness.** The team commander is ultimately responsible for the route the team takes, however the immediate path is down to the scout. Consequently the scout needs to pick his route with care, bearing in mind the size of the team and the fact that they will be carrying the heavier items of equipment. He should avoid steep banks, dead fall and dense vegetation. After selecting his next bound he should try to memorize the route to enable him to move forward without looking down at the ground. There will be times when this can't be avoided. In this instance he should work with the second man to ensure that the vulnerable arc is covered at all times. Therefore if the scout looks down he should also lower his weapon, the second member of the team or his CM on seeing this will raise his weapon to cover the arc to the front.

0422. **Personal Skills.** The scout must always maintain his weapon in the shoulder and his personal skills, i.e., movement and noise discipline must be of the highest order.

0423. **Action on Contact.** In close country, especially when tracking and pursuing the enemy, the chance of contact with either the enemy, friendly forces or civilians is greatly increased. The scout is going to have to identify at the furthest distance whether the movement he sees to his front is friend or foe. The likelihood of a fratricidal (blue on blue) incident is significantly increased in this environment. The decision he makes as whether or not to engage has to be made in a split second.

0424. **Contact Drills.** A TT engaged in information gathering will conduct a contact drill as a six man fire team drill, as laid out in Pamphlet No. 5, Infantry Company Group Jungle Tactics. A TT engaged in pursuit will conduct a section contact drill, as laid out in Pamphlet No. 5, with the hold line being called when the team is back in rough alignment with the rear two men.

Summary

0425. During the Malaya Emergency and Borneo Confrontation the majority of friendly force casualties were lead scouts. Fortunately these were relatively low due to the relentless training carried out by British forces. The soldiers involved trained to a point of being intimate with their weapons, being able to detect a stoppage by sound or feel alone. This is the level that all infantry soldiers operating in a jungle or close country environment need to achieve. A good lead scout should be able to move silently and have the ability to communicate with the minimum of fuss with his team and detect any presence, be it friendly or foe, without first being seen himself. Through hard, dedicated and realistic training all infantry soldiers should be able to attain this standard.

0426 – 0427. *Reserved.*

SECTION 4. — TRACKER TEAM ROUTINE DRILLS

Introduction

0428. The routine drills used by a TT are based on the patrol drills contained in Pamphlet No. 5, Infantry Company Group Jungle Tactics, however, they will differ to those of a rifle section or a long-range patrol team due to the fact that when following or pursuing sign their chances of contact are that much higher. The movement of the VT element is very slow and all team members must maintain high levels of alertness for long periods of time. On occasions, for example when in pursuit, the TT would be supported by a section or platoon from a rifle company. Their main task whilst on the move would be to provide immediate support then take over from the TT when contact is made. The TT should always seek the safety of a larger friendly force where possible when lying up.

RESTRICTED

Team Movement

0429. **Rate of Advance.** The team will move cautiously and deliberately, covering all round as they move. The speed at which they move can be slow dependent on the skill and experience of the VT and the amount of sign created by the target. A well trained and disciplined target will leave little sign and may use deception tactics or select a route over terrain which is not conducive to leaving good sign. The target may also place snap ambushes and booby traps en-route and may move along topographical features, such as ridgelines, where sign is more difficult to follow but the tracks are generally easier to site ambushes on. The VT and tracking team commander must therefore strike a balance between the requirement to gain on or maintain pace with the enemy, and the requirements for stealth and security. The rate of advance will also vary depending on the terrain and conditions encountered.

0430. **Formations.** The TT will generally move in single file in order to allow the VT to carry out his drills without being foul-tracked by other team members. The VT and CM will track, the other team members are to stay in visual range, with the team commander being the link man, and everyone remaining alert. When necessary the remainder of the team will close up to provide protection for the VT and CM.

0431. **Order of March.** The order of march will frequently change within the team due to the rotation of the lead scout/VT. The order of march (see Fig 6) will be the tracking element (VT followed by the CM), the command element (team commander and RO), then the protection element (the next pair of VTs). The CM determines the distance between the lead scout and himself. VTs will be rotated as required. When attached, the dog and handler, when not tracking, are located within the command element.

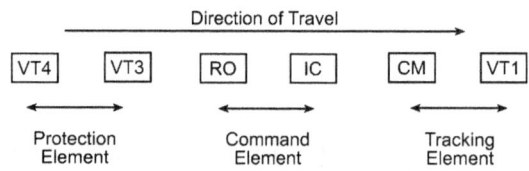

Fig 6. — Tracker Team Order of March

Routine Drills

0432. **Obstacle Crossing Drill.** The TT is extremely vulnerable whilst conducting an obstacle crossing especially in open areas or places that are dominated by high ground. If the enemy is aware that they are being tracked they may leave an ambush party to defeat the TT. An obstacle is an ideal ambush site because the team

RESTRICTED

may well be exposed at this point. In order to cross an obstacle the TT will utilize the caterpillar method as this enables the VT to remain on sign and prevent it from being fouled by other members. The drill is as follows:

    a. VT1 will indicate the obstacle and the remainder of the team will go firm and conduct a listening halt. The VT and CM will then track forward to the entry point and cover the obstacle.

    b. The IC and RO move up to the VT and CM and relieve them covering up and down the obstacle whilst VT3 and VT4 cover the rear. The team commander can control the crossing from a central position.

    c. The VT and CM track the sign across the obstacle ensuring that they confirm the exit point. They will then move, remaining on sign, to the limit of visibility on the far side of the obstacle and then go firm.

    d. VT3 and VT4 then move forward and relieve the IC and RO who subsequently cross the obstacle. On reaching the exit point they go firm and cover up and down the obstacle.

    e. Finally VT3 and VT4 cross the obstacle and rejoin the team. Once complete the commander gives the order to continue tracking.

0433. ***The Tactical Halt.*** The drill is conducted as described in Section 1, Chapter 3, Pamphlet No. 5, Infantry Company Group Jungle Tactics, with the exception of the following:

    a. The VT marks the LDS and back-tracks several metres before breaking track.

    b. On completion of the halt, the team moves to a LUP or back to the LDS to continue "on track" if the light conditions allow.

0434. ***LUP Procedure.*** The TT should remain 'On Sign' for as long as possible. This will normally be from the point after first light when the sign can be identified, up until the point when light has faded and sign can no longer be distinguished. Timings for when sign can be read cannot be dictated and will vary depending upon the terrain, the weather and the ability of the VT. It stands to reason therefore that the TT may have to occupy a LUP sometime before last light and will not be able to commence tracking until sometime after first light. The LUP or overnight halt should follow after the meal/communications halt and will be conducted as follows:

    a. *Occupation.* The occupation of the LUP will be conducted in the same way as for a tactical halt. The team commander must carry out a deadfall check. From here the procedure is as follows: (This differs from what is laid down in Pamphlet No. 5 because of the reasons listed above.)

RESTRICTED

(1) Place out a sentry with claymore mines as required and clear sleeping locations by tying all saplings back.

(2) Stand-to from 15 minutes before until 15 minutes after last light then erect hammocks and shelters, conduct wet and dry routine then rest for the night.

(3) During the night the team must not use any lights or make any noise. The requirement to post sentries will be a decision made by the commander on the ground. He must strike a balance between ensuring that his team remains operationally efficient through being well rested, and the need for security at night. When used sentries are posted at the edge of the LUP.

(4) The patrol commander has to debrief the team as to the track picture over the last 24 hours.

b. *Extraction.* Extraction from the LUP will normally be conducted as follows:

(1) All equipment is packed prior to first light and stand-to.

(2) Stand-to from 15 minutes before until 15 minutes after first light. Prior to completion of stand-to the sentry and claymores are re-deployed out to the limit of noise and visibility.

(3) The team then conducts a meal/communications halt from this location because visual tracking will not be possible until sometime after first light. This is not common with normal section routine drills because they will move away from the LUP some distance before eating. The TT commander may decide to move away from this location. However, he may lose valuable time on track if he does.

(4) On completion of the meal and communications the team moves back onto the LDS to continue on tracks if the light conditions allow.

Change Over Procedure

0435. The VT cannot track indefinitely and must not track when tired. There is no hard and fast rule as to when the VT must hand-over and sometimes handing over may be tactically or professionally unsound, i.e., on arrival at an incident site, or if the VT is still fresh. Flexibility therefore is required. It is essential that when the VT hands over to another VT that the track is not lost, and that security is maintained. In simple terms the VT will hand-over to the CM then become VT4, VT3 will take over as CM and VT4 will move forward one place to become VT3. It is not recommended to swap the VTs in pairs because despite the fact that both are fresh, neither will be tuned in or will be fully aware of the track picture. In detail the drill is conducted as follows:

RESTRICTED

a. *VT to CM.* When the VT states that he needs a rest or is told to change over by the team commander, the VT will go firm and the CM will close up with him. The team commander will cover them. The VT briefs the CM as follows:

(1) The VT shows the CM the sign he is on, which should be conclusive, and indicates his furthest sign.

(2) The VT informs the CM what key sign he has been following.

(3) The VT updates the CM on the track picture, i.e., any information that the new VT may not be aware of, anticipated incidents, etc.

b. *Hand-over.* The CM then tunes in and observes the sign. If he is content that he is on conclusive sign he will take over. However, if he is not content the VT will recommence on track until his next conclusive sign, where the procedure is repeated until the CM is confident that he is on track. The VT reconfirms all of the information that he previously briefed the CM on and steps off track. The CM tunes in and is now ready to continue as the VT. It is important in maintaining patrol security that the CM continues his duties until he is ready to take over as the VT.

c. *The New CM.* The VT3 will move forward ahead of the team commander to cover the new VT whilst he tunes in. He will be fully briefed by either the team commander or the old VT. He then continues on task until it is his turn to become the VT.

d. *Remaining VTs.* The original VT will become last man and VT4 will step up one place. This ensures that each VT has an equal amount of rest and that the task of being the CM is spread equally amongst the VTs and that the CM is always fresh.

Summary

0436. Tracking is an extremely difficult task especially when in close proximity to the enemy. It can be very slow and tiring and it is vital that the team carries out these drills proficiently as this will ensure that their security is maintained.

0437 – 0438. *Reserved.*

RESTRICTED

SECTION 5. — THE PURSUIT

Introduction

0439.   One of the problems a commander has is following the enemy after a sighting or contact has been made. This can be achieved by employing a TT to pursue the enemy and establish contact. The TT (VTs and dogs) evolved during World War II and were reintroduced during jungle campaigns in Malaya, Kenya, Cyprus, Borneo and Vietnam.

Pursuit Tracking

0440.   A pursuit can be conducted in three ways. These are as follows:

    a.   TT with tracker dog attached. This is the most effective grouping.

    b.   TT with VTs only.

    c.   Attaching a tracker dog and handler to an infantry platoon.

Tasking a TT for a Pursuit

0441.   *Reactive Deployment.*   A TT will always be on standby in order to move quickly in reaction to a contact or sighting. A sense of urgency is essential in order for them to locate fresh sign and to prevent them losing too much distance to the enemy. Therefore some detailed planning will be deleted. It is imperative that the force on the ground restricts its movement so that the sign is preserved. After a contact the following action should happen:

    a.   *Warning Order/Briefing.*   Before deployment the TT must be briefed on the incident/contact in outline. The briefing will include the following:

        (1)   A description/sitrep on the incident.

        (2)   Drop off point, RV and method of transport.

        (3)   Unit and C/S of who to liaise with on arrival.

        (4)   Radio details/CEIs/BATCO — if not already issued.

        (5)   Position of all friendly forces in the area.

        (6)   Details of Support Group if known at this stage.

    b.   *Action on Arrival at the Incident Point.*   The team will deploy by whatever means are available. On arrival at the incident/contact point the team will move

RESTRICTED

into all round defence and the signaller will establish communications and confirm any communications arrangements with the support group. The commander on the ground must brief the TT commander as follows:

    (1)   Confirm the present location.

    (2)   Relay all details on the enemy, i.e., time last seen, numbers, weapons, direction of movement, etc.

    (3)   Confirm the known disposition of all friendly forces in immediate area, and the limit of their movement.

    (4)   Confirm the weather before and after incident.

    (5)   Clarify who the support group is going to be.

    (6)   Any other information as the commander on the ground sees fit. The team commander will then brief his team.

0442. **Planned Deployment.** For specific operations a TT will be deployed as part of the task organization (*see* Chapter 1) and will prepare for the pursuit prior to contact being made. On a pre-planned pursuit less time will be spent on briefings because details of both friendly and enemy forces will be known. On completion of the operation troops must be held at the LOE and the TT tasked immediately. The team will confirm the limit of friendly force movement and any immediate details on the enemy before casting for sign.

0443. **Casting for Sign.** The TT will then move to the limit of local movement, and begin casting for sign, ensuring that all friendly forces are informed of their movement. The VT and CM or dog and CM cast for sign whilst the remainder form a baseline as protection to the rear of the tracking element. Once sign has been located the information gained is compared with that obtained on arrival, then the signaller will inform the higher commander that they are on sign. The team and support unit then lead off pursuing the enemy with either the VT or tracker dog and handler leading.

Pursuing the Enemy

0444. A TT will usually lead the pursuit with the aim being to re-establish contact with the enemy. The TT will be assisted by a support group of any strength depending on the size of the enemy. Whether the pursuit is conducted as a pre-planned operation as part of a follow-up force exploiting success, or is conducted in reaction to a contact or sighting, the tactics and formations employed will be the same.

0445. **Types of Track.** The track pursued is classified depending on the age of sign and estimated proximity of the enemy as follows:

    a.   *Cold Track.*   Cold track is deemed as being sign greater than 24 hours old.

RESTRICTED

RESTRICTED

b. *Warm Track.* Warm track is deemed as being sign less than 24 hours old but greater than two hours old.

c. *Hot Track.* Hot track is deemed as being sign less than two hours old.

0446. **Formations.** The formations used by the TT during the pursuit will be dictated by the ground and the type of track they are pursuing. The TT is under the command of the support group commander. At different times either the tracker dog or VTs will head the team and occasionally on good sign the VTs will be able to track at very high speeds. The load of both the TT and the support group must be borne in mind by whoever is tracking so that a constant speed can be maintained. This is to prevent gaps forming within the TT or support group which can lead to a split team. The formations used during the pursuit are as follows:

a. *Cold Track.* The TT will pursue the enemy with the VT/tracker dog conducting the track pursuit drill (TPD) as at Fig 7. The support group will normally travel a short distance behind the TT in order to both allow the team to track, and to avoid a stop start move. This formation will also be used on difficult terrain, on warm hot track. For added protection the CM and IC can deploy to the flanks to cover the VT when stopping to search for sign.

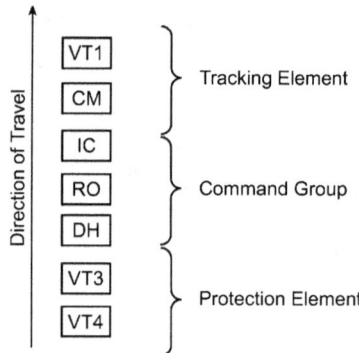

Fig 7. — Cold Track Formation

4-14   **RESTRICTED**

RESTRICTED

b. *Warm Track.* On a warm track with VTs leading the team will adopt the formation in Fig 8. The TPD will not be used. The idea is that a VT will be positioned on either side of the track observing for both sign and the enemy. Each VT will observe for the furthest possible sign and move towards it remaining parallel to the track and will keep moving as long as he has sign on his appropriate flank. If both VTs cannot see sign, for example it may be obscured from one side by the vegetation, they may continue as long as one of the VTs is on sign. On route the other VT may observe further sign and push on to that point. There is no need for talking and if at any time they can no longer see sign they will go firm and search the ground, and if still unsuccessful they will move back to their LDS and start again. It is essential that the VTs do not cross the track. The command element will travel on the track, however must remain a safe distance back in order to allow the trackers to come back to their LDS if required. This is a very fast method of tracking relying upon two men remaining parallel to the sign and affording each other protection.

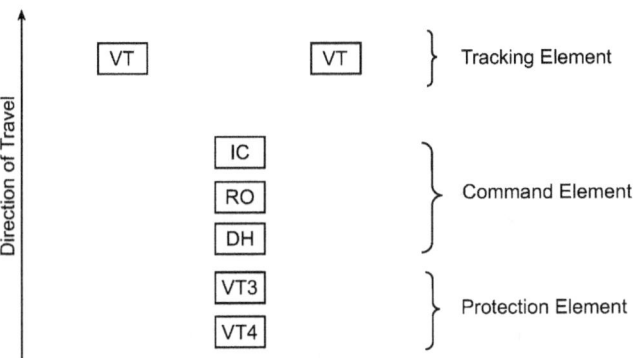

Fig 8. — Warm Track Formation

c. *Hot Track.* The TT will use the formation in Fig 9 on a hot track. The TT commander will use all of his VTs in this formation although the lead VTs will be the principal trackers. The lead VTs will pursue the sign in the same manner as for a warm track. However, in this formation the rear trackers will be travelling parallel to the track behind them. The idea being if at any time the track turns and it is missed by the lead VTs the rear ones will pick it up. The team can then turn in the appropriate direction and continue with the trackers on that side taking the lead. In addition greater protection is afforded by spreading the team out in this manner. This formation is both physically and mentally demanding because all VTs of the TT are tracking and contact with the enemy is more likely. Therefore the support group will be closer behind the TT.

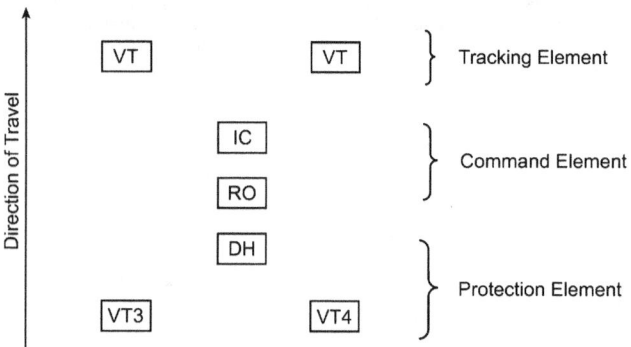

Fig 9. — Hot Track Formation

d. *Formation With a Tracker Dog Leading.* A tracker dog and handler will lead from the front with a CM (*see* Fig 10). Depending on the tactical situation a combination of the above formations may be adopted with the tracker dog still leading. Alternatively the tracker dog may be on a scent trail some distance to a flank off the visual track with the remainder on the track. In any formation the handler must have a dedicated CM. The lead CM is the link-man between the handler and the support group. His responsibilities include:

(1) The passage of information from the handler to the TT support group and vice versa.

(2) Checking areas of interest as instructed by the handler, such as knolls, rivers and streams, etc. This is done in conjunction with the remainder of the TT.

(3) Providing close support for the handler.

RESTRICTED

(4) Taking over as the tracker and relocating the sign as and when required.

(5) Searching any incident site encountered.

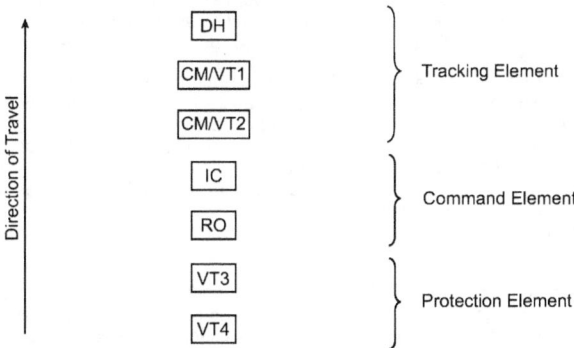

Fig 10. — Formation With a Tracker Dog Leading

0447. **Support Group Location.** The working distance between the TT and the support unit on a cold track is normally 100 metres to ensure that the enemy hears less noise. However, if the situation is such that a contact is imminent, i.e., on a hot track, the supporting unit will be close behind the TT. When the tracker dog is leading, especially when a large enemy force is anticipated, the lead man of the support group will be directly behind the CM.

0448. **Incident Sites.** When a TT arrives at an incident site, it is the responsibility of both the team commander and support group commander to decide how much time to spend at the incident. They must balance the need to gain information from the site with the main aim of closing with the enemy. Once the VT is confident that he has gained sufficient information from an incident site, i.e., he has confirmed the information already known on the enemy he will continue on track. The commander may decide not to spend more than a few moments at future incidents and will literally confirm the directions of entry and exit.

0449. **Action on the Enemy Splitting.** The commander will normally decide to pursue the strongest sign. However, this might not always be the case. For example, when deployed on a pre-planned operation such as a deliberate attack, there will normally be a cordon and intelligence on other likely enemy locations known. On such occasions the commander may decide to pursue an enemy on a route likely to evade the cordon, or pursue an enemy that is not heading towards a known enemy location.

RESTRICTED

RESTRICTED

0450. ***Speed of Movement.*** As stated both the quality of sign pursued and the load of the TT and support group will dictate the speed of movement. The aim of the pursuit is to re-establish contact with the enemy therefore it must be accepted that in order for the pursuing force to catch the enemy he must travel faster than the enemy. On a hot or warm track trackers will go much quicker than on a cold track. As a result the rate of advance will be fast so equipment must be as light as possible and be kept to a minimum. Ammunition, rations and water will be the main requirements.

0451. ***Navigation.*** Navigation will be difficult due to the rate of advance especially on a hot track. The commander must conduct a detailed map study before starting and whilst on the move will need to identify features quickly. The compass will give a direction and pacing will allow the commander to monitor the distance travelled. It is difficult to accurately pinpoint the pursuing forces position, however the commander must endeavour to be as accurate as possible. Main features will be of great assistance and every time the commander stops he should immediately confirm his location. When a cordon is in place the importance of accurate navigation can not be overstated.

0452. ***Cordons and Other Friendly Forces.*** When on a pre-planned pursuit where a cordon is in position the task of the TT and support group is essential. On a deliberate attack there will often be two cordons used, an inner and outer cordon to prevent escape and to prevent reinforcement. However, both of these cordons may not be in place at the time of attack. If an inner cordon only is used the outer cordon will be held in reserve, possibly on another task, and will deploy once the enemy escape routes have been identified. Identifying the escape routes will be the task of the pursuit force and the information provided by them will direct where the outer cordon is inserted and where they set their ambushes. The pursuit force will hopefully drive the enemy into the outer cordon. On such operations it is essential that the TT and support group have the ability to communicate with their higher command.

Action on the Enemy

0453. The task is to find, fix and strike the enemy. By following the enemy in a pursuit the TT is finding the enemy, the support group will fix and strike them. The TT is aiming to relocate the enemy not physically contact them. Ideally they should aim to hand over to the support group just short of the enemy. The reality of the pursuit, however, is that very often the TT will either run into the back of a moving enemy, or will themselves be contacted by a static enemy. The actions employed will be as follows:

    a. *Enemy Pre-Seen/Indicated.* The enemy pursued may be identified for a number of reasons, either having been seen or heard, or most likely through the tracker dog giving an indication. If the enemy is static a scent pool will build in the area and the longer he remains in that location the stronger the scent picture will be. If the enemy is pre-seen the TT will go firm and cover the direction towards the enemy. The support group will then pass through the TT and advance towards the enemy.

RESTRICTED

b. *Action on Contact.* The TT may not have any indication that they have closed with the enemy and may find themselves either walking into a sentry, meeting engagement, or a snap ambush. It is essential that on a warm or hot track that the support group is as close as possible so that they can close with the enemy without delay. In the event of a contact, the following action should take place:

   (1) *Cold Track.* On a cold track when the support group is a tactical bound behind the TT should carry out the contact drill back to the hold line and put down a heavy weight of fire. The commander of the support group will move forward and conduct a quick estimate.

   (2) *Hot or Warm Track.* The support group will be right behind the TT and will attempt to take over from TT as quickly as possible. The TT will carry out the IA drill and immediately the lead section of the support group will join them and take over the firefight. The TT will then withdraw and regroup ready for re-tasking.

   (3) *The Hold Line.* In summary on any contact the TT will remain in the hold line as fire support group until replaced or stood down.

Reorganization

0454. On completion of the assault the support group commander will conduct a reorganization. This will include the following:

   a. Establish a limit of exploitation for his own troops.

   b. Send a sitrep to HQ.

   c. Clear and define the enemy area.

   d. Use the TT to carry out a search of the area and check for sign on likely escape routes.

   e. Make a follow-up plan after the TT has reported the results of its searches. If all of the enemy have not been accounted for tracking restarts.

Summary

0455. The use of a TT with a support group to pursue the enemy is vital in bringing about the destruction of the enemy regardless of whether it is a pre-planned or reactive operation. Tracking in the pursuit is one of the most demanding and dangerous tasks that a tracker can do. The aim of the TT is to find the enemy for the support group so they can re-establish contact. In order to do this it is vital that no time is lost in reaching the scene of the incident so that the track and scent are as fresh

RESTRICTED

as possible. Whilst speed is of the essence in the pursuit, it must also be balanced by the need to ensure security. The speed of hand-over from the TT to the support group is essential if the enemy is to be defeated. To remain at peak performance the TT needs regular training to improve its tracking techniques and experience, thus increasing its chances of success.

ANNEX A TO
CHAPTER 4

TRACKER TEAM KIT AND EQUIPMENT

1. The TT should travel as light as possible however the team commander must ensure that his team has all of the equipment required for his team action. The team should carry all of the equipment normally carried for the environment they are operating in. Equipment should be carried as follows:

   a. *Dress.* As current unit SOPs.

   b. *Equipment.* Personal and team equipment should be minimised as much as possible (*see* paragraph 2).

   c. *Weapons.* Rifles may be carried with or without a sling. A sling will be carried either in a pouch or on the weapon. If it is fitted to the weapon it should be worn in the proper manner. The lead tracker may carry a shotgun. Full scales of ammunition must be carried including grenades and RGGS because of the likelihood of contact. Claymores are also to be carried. The dog handler may carry a sidearm and not a rifle.

   d. *Rations.* It is essential that the team carries rations that can, when required, be eaten cold due to the chances of being in close proximity to the enemy.

   e. *Medical.* One of the team members will be designated as the medic and will carry the team medical pack.

   f. *Emergency Stores.* Ground to air marking equipment, cutting tools and emergency beacons are to be carried (*see* paragraph 2).

   g. *Signals Equipment.* The team signaller is responsible for ensuring that the correct communications equipment is carried relevant to the area of operations.

   h. *Specialist Equipment.* Specialist equipment such as night viewing aids, CTR kit and FUP marking equipment should be carried for use as required.

2. **Suggested Carriage of Team Ammunition and Equipment.**

   a. *VT1/Lead Scout.*

   (1) Radio bty.

   (2) Folding saw.

RESTRICTED

    (3) Claymore.

    (4) 2 packs 1.5V btys.

b. *VT2/Coverman.*

    (1) Waterbag.

    (2) Secateurs.

    (3) CWS.

    (4) 2 packs 1.5V btys.

c. *Commander.*

    (1) PRC 349 complete (or equivalent).

    (2) Binoculars.

    (3) Pace counter.

    (4) Hand generator.

    (5) TAM.

    (6) Strobe light.

    (7) 2 packs 1.5V btys.

d. *Radio Operator.*

    (1) PRC 320 complete (or equivalent).

    (2) Codes.

    (3) 2 packs 1.5V btys.

e. *VT3/2IC.*

    (1) Heli signalling equip.

    (2) Radio bty.

    (3) Waterbag.

    (4) CWS.

    (5) 2 packs 1.5V btys.

RESTRICTED

    f. *VT4/Medic.*

        (1) Medical pack.

        (2) Lightweight stretcher.

        (3) Shovel.

        (4) Folding saw.

        (5) Claymore.

        (6) 2 packs 1.5V btys.

    g. *Dog Handler.*

        (1) Dog rations.

        (2) Water.

        (3) Veterinary first aid kit.

    h. *Ammunition.* The following scales of ammunition are carried by each individual:

        (1) 4 magazines of 5.56 mm SAA.

        (2) 2 bandoliers of 5.56 mm SAA.

        (3) 2 HE grenades.

        (4) 1 smoke grenade.

3. In summary the team must ensure that they are equipped to task. Resupply can be conducted by a variety of means, i.e., jungle line resupply and pre-laid caches as and when required. Depending on the tactical situation especially during pursuit tracking it may be advisable to employ a light/medium machine gun in the team.

RESTRICTED

RESTRICTED

Chapter 5

TRACKING REPORTING

SECTION 1. — GENERAL

Introduction

Contents	
	Page
SECTION 1. — GENERAL	5–1
SECTION 2. — SIGNALS REPORTS	5–2
SECTION 3. — HOT DEBRIEF	5–3
SECTION 4. — VISUAL TRACKING REPORT	5–7
SECTION 5. — TRACKING PRESENTATION	5–15
Annexes:	
A. Signals Report Format	5A-1
B. Hot Debrief Format	5B-1
C. Visual Tracking Patrol Report Format	5C-1

0501. In order for trackers to fulfil their potential as an intelligence gathering asset the information gained by a TT must be reported, collated and acted upon in a timely and efficient manner. On tasking any TT they must be given a mission stating clearly a desired end state, i.e., to gather information on the enemy in order to conduct subsequent offensive operations against them. Subsequent offensive operations based upon the information provided by trackers will normally be in the form of either an ambush or a deliberate attack. Ambushes will be set on enemy routes and cache locations and attacks will be mounted against enemy camps. The success of any follow-up action will be affected greatly by the intelligence upon which it is planned.

0502. TTs will always gain information. However, it will only assist in bringing about successful operations if it is handled in the correct manner. Speed will often be important in seizing the initiative therefore TTs will send reports by radio and must be debriefed as soon as possible following a patrol. TTs will report the information they have gained in the following ways:

    a. Signals reports.

    b. Hot debrief.

    c. Visual tracking report.

    d. Tracking presentation.

0503 – 0504. *Reserved.*

RESTRICTED

SECTION 2. — SIGNALS REPORTS

Introduction

0505. The ability to communicate by radio is crucial to a TT. Radio communications will enable the team to relay information without returning to a friendly force, and to call support whenever they need it. The TT will normally use location schedules sending a sitrep twice per day. Other information of an urgent nature must be passed without delay. The team signaller must be proficient at both HF and VHF communications.

Report Formats

0506. ***The Tracking Report.*** The signals tracking report is used at the cast site or when sign is first located, and should also be used to send any other information as and when required i.e., as a daily situation report or critical information gained on track. The format for this report can be found at Annex A. Not all of the serials have to be sent, especially if the serial contains information that has already been sent or where there is no change. Alternatively "Nil" or "No Change" may be sent during difficult communications, which will keep the report running in sequence. The information sent is a summary of the information gained. If available the following information is sent:

    a.   The DTG of incident/discovery.

    b.   The grid reference of the incident, or point where sign was first located.

    c.   Your location.

    d.   The estimated age of sign at the incident site.

    e.   The general direction the enemy is travelling although more than one direction may be reported.

    f.   The estimated or known enemy strength.

    g.   The possible enemy objective if it can be predicted by the TT. This will also be looked at closely by commanders/intelligence cells and can be predicted more accurately as the track picture develops.

    h.   The possible enemy location at the current time. This may be a grid reference, a grid square or a distance and direction. Commanders and intelligence cells will also consider this in detail and will become easier to anticipate with a more detailed track picture.

RESTRICTED

i. Tactics identified throughout the day, i.e., a brief summary of what tactics they use. The type of incident, as recognising the first incident may help the tracker start to form a track picture especially if he knows the enemy daily routine, i.e., when during the day they would conduct this kind of activity.

j. Your own intentions as a TT. This may be to continue on sign or depending upon the time of day/situation to abandon track, lie up and start tracking the following day.

k. Any other information found that is relevant/critical at this stage (*see* Section 4 in this chapter).

Summary

0507. The signals reports must be sent at the correct times so that the intelligence can be collated as it becomes available. The team will send a tracking report on initially locating sign and then in accordance to the allotted radio schedules.

0508 – 0509. *Reserved.*

SECTION 3. — HOT DEBRIEF

Introduction

0510. The hot debrief is utilised on completion of a patrol as a means of reporting the main information gained. For a TT the hot debrief will normally be delivered to the team's platoon commander, IO, Ops Officer or if attached to a company the company commander or 2IC. Prior to the hot debrief the team commander will debrief his team and collate all of the relevant information. The hot debrief should be conducted no later than one hour of the team returning and the commander must ensure that he concentrates on passing the key information. If possible the complete team should be present during the debriefing. An example of the hot debrief format can be found at Annex B to this chapter.

Hot Debrief Content

0511. ***Preliminaries.*** Preliminaries are used prior to the hot debrief as with any set of orders or briefing. The team commander should always use an aide-mémoire and notebook and pencil so that the debrief follows a logical sequence. The team commander must always use a map or model to illustrate the target actions. The prelims will include the following information, however if the information is known to the person being briefed it can be left out:

a. Call signs and tasks.

RESTRICTED

    b. FL and LL.

    c. DTG in and out.

    d. Map coordinates.

    e. Climatic conditions (per day).

    f. *Model/Sketch/Map Description.* The commander should produce a patrol trace/model/sketch, which he will use to describe the route taken, and identify incidents.

    g. *Ground Orientation.* An overview of the area of operations.

0512. **Situation.** Only to be given if the person receiving the brief requires it.

0513. **Mission.** What it was.

0514. **Insertion.** A brief overview of the method, location and the route to the cast site. The remainder of the brief should follow as a clear, concise and chronological story of the vital information gained on track.

0515. **Initial Cast.** The cast site should be briefly described in order to establish an initial intelligence picture for the person being briefed. The following information should be given:

    a. The DTG the sign was located.

    b. The grid reference of the cast site or where the sign was first located. This should also be pointed out on the map/model.

    c. Entry and exits to the cast area.

    d. Direction of travel into and from the cast site, i.e., the direction that the sign was pursued, this should be given as a cardinal point.

    e. The numbers identified at the cast site. This may well change as the track picture is built up.

    f. Age of sign at the time of discovery.

    g. Any other information that you feel is relevant.

0516. **Route Taken and Incidents on Track.** During the hot debrief there is no requirement to talk through each day in detail. Instead, using the patrol trace, the team commander should state the following:

    a. Establish what day he is briefing on, i.e., patrol day 1, and confirm the date of that day. This will make it easier for the person being briefed to understand the patrol action.

RESTRICTED

RESTRICTED

    b. Show the route tracked, breaking it down into the ground tracked each day illustrating on the patrol trace where incidents were located. The patrol trace must be marked accordingly.

    c. Talk through briefly any major incidents encountered however if a similar incident has already been discussed on a previous day state the type of incident located then leave it. Incidents should be described very quickly as in state the method of occupation, the tactics used, i.e., all round defence, what type of incident it was.

    d. Any critical information gained.

    e. Any key information gained that day.

    f. The time, location that the track was abandoned and distance covered each day

    g. The age of sign at the start and end of each day.

0517. *Patrol Summary.* This is the important part of the hot debrief and contains the detail required by commanders to plan and conduct offensive operations based upon the information provided by the TT. The patrol summary delivered during the hot debrief must be short and to the point concentrating on the main issues whilst allowing the person being briefed to ask questions. The following information is given:

    a. Total distance tracked.

    b. Total time on track.

    c. Numbers being tracked.

    d. Sex.

    e. The enemy's direction of travel.

    f. *Speed and Load.* The speed that the target is travelling at in conjunction with the load he is carrying, i.e., 2 km per day carrying packs of 18 – 22kg.

    g. General health.

    h. *Habits and Routines.* Describe the enemy's habits and routines, i.e., his sequence of events during his patrolling day.

    i. Rations.

    j. Equipment.

    k. *Weapons.* If identified on track.

RESTRICTED

l. *Tactics.* The tactics identified and an assessment of their tactical awareness is essential for planning further operations. For the purposes of the hot debrief the assessment of tactical awareness is more significant because this should be used to summarise the enemy's tactical ability.

m. *Critical Information.* This will vary during the patrol, the first time a weapon is identified will be critical but the next identification will not as the fact they have weapons has been reported.

n. Other information.

0518. **Conclusions.** Your conclusions or assumptions on the track picture you have gained should include the following:

a. The type of target whether they are regular troops, SF, part time, local indigenous, etc. It should also cover the type of action they were carrying out, i.e., fighting patrol, reconnaissance patrol, administrative move, etc.

b. Tracker awareness/deception employed. Was the target aware that they were being followed. If deception tactics were employed was it as a result of your patrol or did they conduct them as a mater of routine.

c. *Possible/Predicted Intentions.* The possible objective or intentions that you assume the quarry is endeavouring to undertake. Be prepared to quantify your assumption with evidence.

d. The possible or predicted location is given in addition to the possible or known objective in order to identify where the enemy is if he is expected to be still on route. Alternatively if the objective is not known a predicted enemy location will prove useful for tasking future patrols.

0519. **Recommendations.** Your recommendations as to what follow-up action should be taken against the enemy.

Summary

0520. The hot debrief is a vital tool for passing information first hand without wasting time. In essence the commander or int/ops rep will debrief the patrol whilst the team commander will be briefing him on what he has discovered. The hot debrief must deliver the important information and allow the detail to be passed in a full tracking report. It is vital that a full tracking report is produced and read as it will inevitably prove to be a far more comprehensive and useful aid when planning subsequent operations against the enemy.

0521 – 0522. *Reserved.*

SECTION 4. — VISUAL TRACKING REPORT

Introduction

0523. The signals reports and the hot debrief are fundamental in reporting the information gained from the track. However this information will often be collected by more than one agency. The hot debrief in particular will probably not be written down and recorded except in the patrol notebook. This information over a period of time may become mislaid and intelligence will be lost. The visual tracking report is the primary means for recording information. When produced the final product will include a full record of the information gained on the enemy. Furthermore the visual tracking report will be archived and recorded as part of the overall operational intelligence picture and can be used to inform planners of future operations.

Visual Tracking Report Format

0524. The visual tracking report is produced by the whole TT especially if the patrol has been of a long duration and a number of incidents identified. VTs will record incidents encountered whilst the team commander records his findings, summary and recommendations. Whilst tracking it is common for the VT and CM to record incidents encountered in their notebooks which are subsequently rewritten into a dedicated patrol notebook in the LUP. On completion of the patrol the team will use the information contained in the patrol notebook, also known as the patrol log, to compile the report. The report format can be found at Annex C.

0525. **Preliminaries.** Preliminaries are used to record the patrol designation and composition, indicate the area of operations, a general outline of the ground covered, the climatic conditions on a daily basis and the patrol's mission. The prelims are allowed on the report by a summary of the situation at the time the patrol was deployed and patrol's mission as follows:

Mission. The mission statement as given to the patrol is recorded so anyone reviewing the report can see straight away what the patrol had been tasked with.

0526. **Patrol Details.** The patrol details part of the report records the patrol duration, movement details, distance tracked and the period on track. The period deployed will be longer than the time on track. There will therefore be a delay, depending upon the method of extraction, i.e., how long it takes to return to the unit receiving the report, between when the information was gained and when it was received. It is advisable that in order to avoid too long a delay TTs be withdrawn taking into account the tactical situation by the fastest means available. The distance and period spent on track are important information as they will illustrate the period, in both duration and length of track, over which the information within the report was gained. An example of how to complete the patrol details box is at Fig 11:

RESTRICTED

Patrol Duration:	Movement Details:	Time on Track:
Departure DTG: *020700 Sep 00*	Method of Insertion: *Heli Abseil*	Start DTG: *021430 Sep 00*
Return DTG: *070700 Sep 00*	DOP GR: *102 583*	Abandon track TDG: *061700 Sep 00*
	Method of Extraction: *Boat*	Start track GR: *24681234*
	PUP GR: *194 565*	Abandon track GR: *19421234*
Total Period: *5 days (120 hrs)*	Distance Tracked: *11.2 km*	Total Period Tracked: *45.5 hrs*
		Total Distance on Track: *9.5 km*

Fig 11. — An Example of a Completed Patrol Details Box

0527. ***Cast Site Details.*** The cast site details box (*see* Fig 12) is used to record all details of the cast site including the Grid reference and DTG of the sign that was located. The report also includes a description of the area, i.e., a topographical description of the area, and the targets entry and exit bearings. A sketch is made of the incident site and as with all other sketches a scale and North pointer are used. This sketch will be recorded in the patrol log and, if needed, be reproduced as an annex to the report. The information gained at the cast site, in accordance with the information gained from tracking, is recorded as follows:

    a. *Entry and Exit Bearing.* If it can be assessed that the bearings are part of a deception, due to circumnavigating an obstacle or because of a topographical feature then the assessed general direction should be recorded as a cardinal point in the other information box.

    b. *Numbers Identified.* This may well be an estimate as in a bracketed figure and will be confirmed as the track picture is developed.

    c. *Age of Sign.* The age of the sign found at the cast site.

    d. *Other Information.* This box is used as required.

RESTRICTED

Grid Reference: *0910 5780*	DTG: *021430 Sep 00.*
Description of Area: *The incident location was a ridgeline track on the top of a knoll, following a climb of around 250 m.*	
Entry Bearing: *5800 mils*	Exit Bearing: *3200 mils*
Numbers identified: *5-7 persons*	Age of sign: *12-18 hrs*
Other information: *The quarry conducted a tactical pause and the sign of one bipod mounted weapon was located. General direction of travel South-West.*	

Fig 12. — An Example of a Completed Cast Site Details Box

0528. **The Route Followed.** (*See* Fig 13) This is the main body of the report and must be recorded as accurately as possible. This will most likely be a direct reflection of the patrol log and it is imperative that each incident is recorded in chronological order. It starts with the insertion and ends on the extraction. The DTG and the reason the patrol abandoned track must also be included. This part of the report should be compiled as follows:

    a. *DTG.* The first DTG should be the DTG the patrol inserted, the remainder should then run in order.

    b. *From.* The point reflected in the DTG and expressed as an eight figure grid reference.

    c. *To.* A significant part of the track picture, depicted by incidents conducted by the quarry or topographical features that form a natural break in the track picture, i.e., A spot height, a change in quarry direction, TGT LUP/short halt, etc. These should be expressed as a grid reference.

    d. *Direction.* May be expressed as a cardinal point. This is not a leg on a route card so each minor change in bearing need not be recorded.

    e. *Distance.* Expressed in kilometres or metres.

    f. *Age.* Recorded as a bracket, i.e., 12 – 18 hrs, 2 – 6 hrs, etc.

    g. *Facts.* This is an accurate record of any facts that have been located on the length of track recorded between the From and To columns, i.e., Four areas of flattening, cigarette end, rubbing marks on trees, flannelette 100 mm x 50 mm, etc.

Route followed: Example

DTG	From	To	DN	Dist	Age	Facts	Assumptions
120800	96206820	95806752	SW	900m		Insertion and transit.	
120830	95806752				12-18 hrs	Cast site. 4 areas of flattening, cigarette end, boot print (see Annex B1).	A tactical pause/nav check en route, one person smokes B&H US brand, one person wears vibram type pattern boot size 8-9. 4 persons being tracked.
120900	95806752	95256745	West	500m	12-18 hrs	Moved along ridge.	
120950	95256745	95376700	South	625m	12-18 hrs	Moved along ridge. Vibram boot print located size 9-11. Marks on saplings.	They are carrying packs/webbing possibly rifles.
121125	95376700	95306694	SW	75m	12-18 hrs	Broke track 90° after 25m 5 areas of flattening located. After 75m 6 areas of flattening located. Rubbing marks on trees. Large sapling bent down in a North Dn with string fibres in the upper branches. Remains of a small fire at two of the areas of flattening with the remains of noodles and rice. 15m west a latrine was found.	The tgt broke track and carried out a snap ambush then moved into dead ground to conduct a LUP. There are 5-6 persons who cooked a meal in twos/threes eating instant noodles and cooking long grain rice. They used small twigs for fuel. Hammocks and ponchos were erected. One person has loose stools. They are using HF comms and are transmitting to the north.
121205	95306694	95266684	SW	100m	2-6 hrs	Boot prints in the sand. 4 areas of flattening, bipod marks located.	A tactical river crossing drill was carried out. Protection on the near then far banks. A heel count confirms 6 persons, one size 4-6 pattern unknown (see Annex B2) possibly a female/Asian/juvenile. They have one LMG/bipod rifle.
121240	95266684	94646669	West	550	2-4 hrs	Moved along the stream in the water for the duration, then crossed the border.	Possibly tracker aware.
121430	94646669					Abandoned track.	Due to tgt crossing an international border.

Fig 13. — An Example of a Route Followed

h. *Assumptions.* This is your interpretations of the facts and your assumption of what the target has done at this point, i.e., a 7.62 mm weapon was pulled through, one person smoked a cigarette, a poncho was erected, the quarry conducted a short meal halt/LUP, etc. Also any new information gained should be recorded, i.e:

(1) Tactics employed as in the method of occupation as this will help to build an understanding of the enemy's modus operandi and assist not only TTs looking for sign but friendly forces who may contact the enemy.

(2) Numbers identified.

(3) Any sign of weapons.

(4) Discardables should be retrieved to be processed by the int cell on completion of the patrol especially unidentifiable, rare or critical items. Multiple finds of the same item such as foodstuffs may be left at the incident site. **Trackers must always remain booby trap aware when they find such items.**

(5) *Critical Information.* Any information discovered that indicates either the enemy mission or objective, a threat to friendly forces or identifies further enemy should be deemed as critical information. Booby trap markers, saplings bent over indicating the direction of enemy communications, identifying an enemy RV where patrols have come together are all examples of critical information. This would have been immediately sent on discovery, delivered on the hot debrief but it must also be recorded here.

(6) *Any Other Information.* Any other piece of information should also be recorded irrespective of how insignificant you feel it is. It may be critical information to the report reader.

0529. **Overall Target Information.** This is a collation of all the information gained over the whole track picture. Every detail located regarding the quarry should be recorded as follows:

a. *Numbers.* The total number of quarry tracked. This may have to be broken down to explain a patrol split or a tgt RV, i.e., four persons were tracked to grid 12345678 where sign of a further four persons joined the track.

b. *Sex.* If known or assumed must be recorded, i.e., three males and one female. Or, as in the example above, the sign of a female was identified after the target RV. This is not to say the female was in the joining patrol, it may be that she was in the original four but no sign was located.

c. *General Direction.* The over all direction that the quarry was heading.

d. *Speed and Load.* The speed that the target is travelling at in conjunction with the load he is carrying, i.e., 2 km per day carrying packs of 18 – 22 kg. This may be broken down into distinct sections of the track picture.

e. *General Health.* This is the overall state of the patrol and may well be identified on route at various incidents by examining latrines and rubbish pits, i.e., loos, stools, bandages, type of food consumed, etc.

f. *Habits and Routines.* Any recurring actions carried out by the quarry other than tactics.

h. *Rations.* The analysis of discarded foodstuffs may identify the supplier or a country of origin of the target.

i. *Equipment.* Any equipment that you know or assume them to be using.

j. *Tactics.* The tactics identified and an assessment of their tactical awareness is essential for planning further operations. Any tactics that are assumed to have been carried out must be recorded.

k. *Possible Objective.* If identified from the track by either the general direction of travel or information identified. This is critical information.

l. *Predicted/Current Location.* Taking into account the age of sign, direction of travel and the enemy's speed of movement, the location where the TT predicts the enemy will be at this time. This should be expressed with a DTG, i.e., possible enemy location map coordinates E94-93 N66-68 DTG 121600 or possible en location as at 121600 GR 935664.

m. *Tracker Awareness/Deception Employed.* Was the target aware that they were being followed. If deception tactics were employed was it as a result of your patrol or did they conduct them as a mater of routine.

n. *Tracking Information.* This is recorded in order to provide a record not only for the TT, but teams that may be tasked either against this enemy or in the same area in the future, e.g:

(1) *Key Sign.* What the key sign has been during the tracking period. This will assist future TTs in the area.

(2) *Sign Pattern.* Recorded in case foul track is encountered or track isolation is carried out. This will also be used to identify this enemy in the future. It must be remembered that routine drills, i.e., the way the enemy patrols, will often be the same from patrol to patrol and will not provide a sign pattern unless there is a distinct difference in the quality of troops, i.e., regular to special forces. Idiosyncrasies and certain things identified out of the ordinary will provide a sign pattern.

o. *Other Information.* All information not previously recorded should now come under this heading, i.e., a description of any discardables.

0530. **Conclusion.** Your conclusions or assumptions with regard to what the target has done and is likely to do.

0531. **Recommendations.** Your recommendations as to what follow-up action should be taken against the enemy.

0532. **Debriefing Officer's Remarks.** This is for the debriefing officer to record his observations, assumptions, etc.

0533. **Annexes.** A route overlay is mandatory as this illustrates the enemy route that was followed by the TT. A recognisable key should be used and where track isolation has been used the assumed enemy track should be marked differently to the track actually followed. A Going report is also mandatory. A sketch should be included to clarify information where needed. A sketch of an incident encountered should be recorded (*see* Fig 14), or to illustrate the tactics the quarry employed.

Summary

0534. The visual tracking report will be used as a source document by any friendly force commander tasked with the offensive follow-up onto the enemy. The credibility of the TT will be a direct reflection on how the information gained is portrayed in the report, the quality of which will inevitably effect whether any follow-up is successful or not. Time and care must be taken in compiling the report. The team commander and 2IC must proof read it before submitting it.

0535 – 0536. *Reserved.*

RESTRICTED

Incident Encountered Details: Example

Appendix 16 to
Annex B to
Visual Tracking Report
C/S - - - - DTG - - - - - -

a. Incident No *1* Of *16*.	b. Grid Reference: *1095 5927*	c. DTG: *021640 Sep 00*
d. Description of Area: *The incident was located in primary jungle off of a ridgeline track. The area where the enemy went into all round defence was in dead ground to the track.*	e. Entry Bearing: *2600 mils* f. Exit Bearing: *6350 mils*	

g. Sketch:

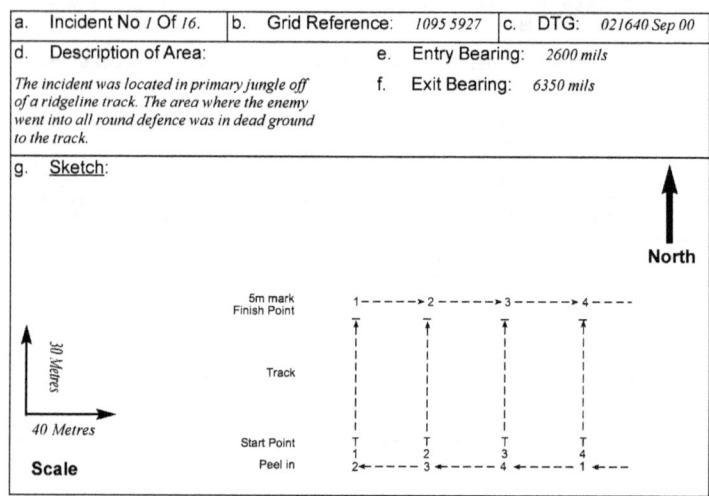

Incident Description:

h. In Outline: (Summary of incident) *This incident was a tactical halt. The enemy broke track and placed a snap ambush on the track using deadfall as cover. They then moved further back from the track into a possible second ambush position, the enemy then moved back into all round defence. They placed a sentry out and on extraction moved parallel to the track before moving back onto it.*

i. In Detail: (Information Gained)		
(1) Method of Occupation: *Double fishhook and snap ambush.*	(2) Numbers Identified: *4 possibly 5 enemy. 4 in the ambush and all round defence plus a sentry.*	(3) Age of Sign: *24 – 36 hours old.*
(4) Tactics Employed: *Snap ambush on breaking track, all round defence and the use of a sentry.*	(5) Types of Weapons: *No sign of weapons identified.*	(6) Discardables: *None found.*
(7) Critical Information: *None other than tactics employed.*	(8) Any Other Information: *The enemy used deadfall to provide cover from fire whilst in their snap ambush location.*	

Fig 14. — An Example of an Incident Encountered

RESTRICTED

SECTION 5. — TRACKING PRESENTATION

Introduction

0537. The tracking presentation is a tool used as a means of formally reporting the information gained to friendly force commanders. The tracking presentation may be used to report all the information gained by the patrol, or for specific operational purposes as little as a single incident. Trackers will be trained in how to deliver a formal presentation as it is a very good training aid for confirming the information they have gained. Many forces use the presentation as the primary means of reporting where time and the operational scenario permit.

Prelims

0538. **Visual Aids.** The team should utilise whatever they have available to them in order to give a clear visual presentation. Visual aids should be used to illustrate the following:

    a. *The Route Followed.* The route followed can be illustrated by using one or a combination of:

        (1) *A Model.* Depending on the distance tracked a model may be built covering either the whole or part of the route. Markers should be used accordingly to identify incidents, LUPs, PUP, DOP, etc.

        (2) *Map and Overlays.* A map with overlays showing the route taken is particularly useful if a large area needs to be covered.

    b. *Incident Sites.* The information gained at incident sites can be illustrated by using the following techniques:

        (1) *Sketches.* A large sketch of the incident drawn as a copy of the one depicted in the track report. A scale, North pointer and incident details such as grid reference and DTG should be on each sketch.

        (2) *Incident Model.* A model of an incident is easily prepared. This method is less time consuming than drawing sketches.

        (3) *Evidence Boards.* An evidence board is normally a piece of cardboard with the discardables located at an incident site attached. There is no requirement to show foodstuffs or easily recognisable objects but all unusual or uncommon objects must be included. The board must include a heading, DTG and a GR.

(4) *Tracking Symbols Card.* A large legend should be visible to all so that the information included on either the sketch or model can be easily understood.

Method of Presentation

0539. **Patrol Introduction.** The TT commander will deliver the introduction. He will introduce himself and his team then describe his visual aids. The team then presents the information gained following the sequence of the visual tracking report.

0540. **Responsibilities.** The tracking presentation is delivered by all and is very much a team effort. The team commander leads the brief when describing the route and the information gained along the route. The VT describes incidents on track at the time the team found the incident site. The commander delivers the patrol summary.

Summary

0541. The tracking presentation is an excellent briefing tool or training aid that can be delivered when required. The commander may present information on a specific incident on completion of a patrol. Alternatively the tracking presentation may be given to soldiers unfamiliar to the area.

RESTRICTED

**ANNEX A TO
CHAPTER 5**

SIGNALS REPORTS

1. ***Tracking Report.*** The tracking report is sent using the format below. Information can be taken from either the tracking vocab card or any other vocab cards.

TRACKING REPORT — REPORT NO. — 104			
SER	TO:	FROM:	FIGURES
A	DTG OF INCIDENT/DISCOVERY (S)		6 FIGS HRS
B	GRID REFERENCE OF INCIDENT (S)		8 FIGS
C	LOC STAT		8 FIGS
D	AGE OF SIGNS		4 FIGS
E	DIRECTION OF TRAVEL		BOX 2
F	ENEMY STRENGTH		BOX 1
G	POSSIBLE EN OBJECTIVE		BOX 4
H	PREDICTED EN LOCATION		BOX 4
I	TACTICS IDENTIFIED		
J	SPEED & LOAD		BOX 6/7
K	OWN INTENTIONS		
L	ANY OTHER INFORMATION		

Note:
Not all serials have to be sent or send NIL/NO CHANGE.

2. ***Tracking Vocab Card.*** (*See* next page) This card can be used for reporting or tasking purposes.

RESTRICTED 5A-1

RESTRICTED

Tracking Vocab Card

BOX 1		BOX 2	
Enemy Strengths	Code	Directions	Code
Company	11	East	20
Platoon	12	West	21
Section	13	North	22
No. of men (2 figs)	14	South	23
Approx No. of men (2–2 figs)	15	North East	24
Plus strength	16	North West	25
Minus strength	17	South West	26
Possible enemy objective	18	South East	27
Possible enemy location	19	Track Splits	28

BOX 3		BOX 4	
Incidents	Code	The Pursuit	Code
Incident No. (2 figs)	30	Cold track	40
Tactical pause	31	Warm track	41
Short term halt	32	Hot track	42
LUP	33	Enemy direction	43
Live Cache	34	Enemy located out of contact	44
Used Cache	35	From Gd (6 figs) to Gd (6 figs)	45
Booby trap marker	36	At Grid (4 figs)	46
Booby trap	37	At Grid (6 figs)	47
Dead letter box	38	At Grid (8 figs)	48

BOX 5		BOX 6	
Tactics	Code	Tracking Actions	Code
Method of occupation	50	Continuing on track	70
Fishhook	51	Carry/ing out track isolation	71
Broke track using	52	Cast/ing for sign	72
Double fishhook	53	Likely area search Gd (6 figs)	73
Snap ambush	54	Age of sign (4 figs, 2–2 hours)	74
All round defence	55	Foul track encountered	75
Deception encountered	56	Followed by	76
Walking backwards	57	Km per hr (2 figs)	78
Conversion of sign	58	**BOX 7**	
Brushing the track	59	Miscellaneous	Code
Stone hopping	60	Tracker dog	80
Fade out	61	Discardables	81
In a stream	62	Antenna direction facing	82
Splitting up	63	Lightly equipped	83
Walking along a log	64	Heavily equipped	84
Backtracking	65	Change to report (3 figs)	02
		Change to card (3 figs)	03

RESTRICTED

RESTRICTED

ANNEX B TO
CHAPTER 5

HOT DEBRIEF FORMAT

Prelims

1.   **a.**   C/S and TASKS.

    b.   FL/LL.

    c.   DTG in/out.

    d.   Map co-ords.

    e.   Climatic conditions (per day).

    f.   Model/sketch description (use a map).

    g.   Ground orientation.

Situation

2. Brief description of why the task was carried out.

Mission

3. What it was.

        (The above information relayed to an outside agency only)

Insertion

4. Foot, heli, vehicle, boat, etc . . . . . grid and location. The brief from this point should be given clearly and concisely and told as a chronological story.

Initial Cast

5.   **a.**   DTG on track:

    b.   Location (show on map):

    c.   Entry and Exits located:

RESTRICTED

RESTRICTED

    d. Direction of travel:

    e. Numbers:

    f. Age of sign:

    g. Other information:

Route Taken (Day by Day)

6. Explain on model/map route taken.

Incidents on Track

7. Only an overall assumption as to what the incident was and any relevant information. (Taken from serial 9, d – m.)

Track Abandoned

8.   **a.** Time.

    b. Location.

    c. Age of sign.

    d. Distance covered.

(Repeat serials 6 – 8 until the track picture is complete)

Summary of Information

9.   **a.** Total distance tracked.

    b. Total time tracked.

    c. Numbers being tracked.

    d. Sex.

    e. General direction.

    f. Speed and load.

    g. General health.

    h. Habits and routines.

RESTRICTED

- i. Rations.
- j. Equipment.
- k. Weapons.
- l. Tactics.
- m. Critical information.
- n. Other information.

Conclusions

10.
- a. Type of target (regular/SF).
- b. Tracker awareness/deception employed.
- c. Possible/predicted intentions.
- d. Possible/predicted locations.

Recommendations

11.

RESTRICTED

RESTRICTED

ANNEX C TO
CHAPTER 5

VISUAL TRACKING PATROL REPORT FORMAT

Copy No:
PH C/S:
DTG:

Map Coords

Map Reference:

1. Ground (General):

2. Climatic Conditions:
    a. 48 hrs prior to task:
    b. 24 hrs prior to task:
    c. Day 1:
    d. Day 2:
    e. Day 3:
    f. Day 4:
    g. Day 5:

3. Team Composition:

Number	Rank	Name	Appointment/Remarks

RESTRICTED

5C-1

RESTRICTED

4. Mission/Task.

5. Details.

Duration	Movement	Time and Distance on Track
Departure DTG:	Method of Insertion:	Start Track DTG:
	DOP Grid:	Abandon Track DTG:
Return DTG:	Method of Extraction:	Start Track GR:
	PUP Grid:	Abandon Track GR:
<u>Total Period</u>:	<u>Total Distance</u>:	<u>Total Period Tracked</u>: <u>Total Distance on Track</u>:

6. Cast Site Details.

Grid Ref:	DTG:
Description of area:	
Entry Brg:	Exit Brg:
Numbers identified:	Age of sign:
Other information:	

5C-2 **RESTRICTED**

RESTRICTED

ASSUMPTIONS	
FACTS	
AGE	
DIST	
DIR	
TO	
FROM	
DTG	

RESTRICTED

RESTRICTED

8. Overall Target Information.

    a. Numbers:

    b. Sex:

    c. General Direction:

    d. Speed and Load:

    e. General Health:

    f. Habits/Routines:

    g. Rations:

    h. Equipment:

RESTRICTED

RESTRICTED

i. Weapons:

j. Tactics:

k. Possible Objective:

l. Predicted Current Location:

m. Tracker Awareness/Deception Employed:

n. Tracking Information:

o. Other Information:

RESTRICTED

RESTRICTED

9. Conclusions:

10. Recommendations:

11. Debriefing Officers Remarks:

Debriefing Officers: Team Comd:

Signature: Signature:

DTG: DTG:

Annexes (delete as required):

A. Tgt route overlay.
B. Sketches.
C. Going report.
D. CTR report.
E. Ambush recce report.

RESTRICTED

Chapter 6

TRACKER TRAINING

SECTION 1. — GENERAL

Introduction

0601. Training the tracker is the responsibility of the tracking instructor. The tracking instructor must have passed either the Jungle Warfare Tracking Instructor Course (JWTIC) conducted twice per year by JWW in Brunei, or the NZSAS Tracking Instructor Course. The tracking instructor is qualified to plan and conduct tracker training and should be allocated all of the resources he requires. It will take a minimum of two weeks to teach a soldier the tactics, techniques and procedures associated with visual tracking. It is essential that potential visual trackers possess the qualities listed in Chapter 1.

Contents	
	Page
SECTION 1. — GENERAL	6–1
SECTION 2. — TRAINING THE VISUAL TRACKER	6–1
SECTION 3. — TRAINING THE TRACKER TEAM	6–7
SECTION 4. — TRACK AND INCIDENT LAYING	6–10
SECTION 5. — TRACKING PRESENTATIONS	6–11
Annex:	
A. Memory Training	
B. Track Laying Report	

0602 – 0603. *Reserved.*

SECTION 2. — TRAINING THE VISUAL TRACKER

Resources

0604. The tracking instructor can train any number of VTs. However, if he is to train them to a proficient standard he should not be expected to train any more than four VTs at a time. During the early stages of training the instructor to student ratio should be as low as possible, preferably 1:1 or 1:2. To train a TT the tracking instructor should be allocated the following assets/facilities:

    a. *A Fresh Training Area.* It is essential that VTs are trained in areas where foul track is unlikely to be encountered as this will make the task of the instructor far easier. The area should be as large as possible so that fresh tracks can be laid and pursued for every tracking exercise. The area should also contain different types of terrain including all types of jungle, close country, forest, grass areas and rocky ground.

RESTRICTED

b. *A Dedicated Enemy.* An enemy of up to section strength, preferably commanded by a VT, is an ideal enemy force. It is essential that they are briefed thoroughly by the tracking instructor and that they conduct all of their actions in accordance with the direction given and record all of their actions accurately. The enemy lay tracks including incidents to be pursued by the TTs. The enemy patrols are normally called laying patrols.

c. *Service Support.* Service support assets should be made available as required in accordance with the training programme.

Training The Visual Tracker

0605. The primary consideration during tracker training is training the VT. A VT should be trained in all of the tactics, techniques and procedures contained in Chapters 1 – 5. The training must be progressive and flexible enough to allow more time to be spent on various aspects of training as required. Poor weather may well have adverse effects on the training. There must therefore always be some flexibility in the programme. If a high standard of visual tracking is to be achieved the correct instructional procedure of explanation, demonstration, imitation and practice must be followed.

Sequence of Training

0606. **Introduction to Tracking.** The potential VT must be aware of the duties and tasks of both the VT and TT. He must also have a thorough understanding of the basic terminology used in tracking and should be able to explain each term.

0607. **Elements of Tracking (EOT) — The Fundamental Skills.** Of paramount importance is the training received by the VT on the fundamentals of tracking. Each EOT should be taught and practised as follows:

a. *EOT 1 — Sign.* The VT must be able to recognise each of the characteristics of sign, both top sign and ground sign, and be able to classify sign as being either conclusive or inconclusive. Recognising sign is the key to being able to track and should be taught as follows:

(1) *Initial Instruction.* The instructor should initially teach EOT 1 by the use of a lecture demonstration visually showing examples of each characteristic. A sign lane/stand should be set up prior to the lesson in order to do this.

(2) *Pace to Pace Tracking.* Pace to pace tracking is a skill that should be practised during the early stages of training. The VT will follow one man's sign over various terrain. The VT should attempt to identify as many of the characteristics of sign within each pattern as possible before

RESTRICTED

moving on to the next step. The tracks are laid using the pace to pace tracking box as follows:

> *(a)* Each tracker lines up on a base line 5 – 10 metres apart. They make a mark with their foot in a 'T' shape then place their right foot on the apex of the 'T'.
>
> *(b)* Each tracker then walks on a bearing for a set distance, measured by pacing, and makes another mark at his finish point. They then pace forwards a further five metres and halt.
>
> *(c)* The tracker on the far left then turns right and walks to the tracker to his right, that tracker then leads to the next and so on until they reach the far right hand man. The far right hand man then walks a further 10 metres to the right. He then turns to his right and returns to the baseline on a back bearing.
>
> *(d)* The right hand man then leads the remainder along the baseline and occupies the second mark on the baseline, i.e., one place to his left. Each subsequent tracker then peels around him into the next available 'T' shaped start point apart from the tracker who started on the far left who occupies the first 'T' shape. In the end each tracker has moved one place to their left, with the left hand man now on the right hand side (*see* Fig 15 below).

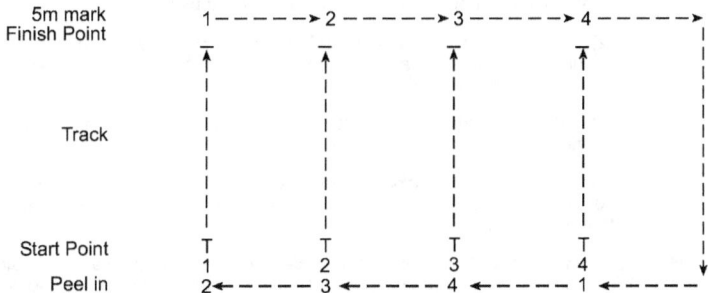

Fig 15. — **Pace to Pace Tracking Box**

b. *EOT 2 — Factors Which Affect Tracking.* The VT should be made aware of these factors early during his training. He will gain a greater understanding of the factors with experience. However, to assist him he should track on a variety of terrain in order to recognise the types of sign left in different environments. In addition he should gain experience of tracking in all weather conditions.

c. *EOT 3 — Judging The Age of Sign.* The ability to judge the age of sign will come with practice and experience. Following the initial lesson the VT should learn to judge the age of sign as follows:

(1) *Age Stands.* Each VT should construct an age stand as a means of learning to judge the age of sign. He should construct a stand similar to a model pit approximately 2 metres x 2 metres in size. He should then place several materials in it and record the date, description and appearance of each material, i.e., a broken twig, a leaf with a straight edge, a fern with a fold, etc. He should then place the same materials into the stand next to their predecessors at the start and end of each day and compare each material to its predecessor. He then records the same details as above, and in addition the changes in appearance of each of the previous materials. Over a period of time he will be able to recognise the age of sign more readily. Key sign for the area should be used.

(2) *Comparison.* The tracker should be encouraged to use comparison of colour as frequently as possible in order to judge the age of sign.

d. *EOT 4 — Information.* The tracker will soon learn what information he can gain from tracking however the greater his knowledge of reporting and presenting this information the more valuable an information gathering asset he will become. Training should include the following:

(1) *Signals Reporting.* Trackers should utilise all of the signals report format and must be proficient at communicating by radio.

(2) *Hot Debriefs.* On completion of each patrol the team commander should be subjected to a full hot debrief. Where possible this should be with the officer tasking the patrol.

(3) *Patrol Reports.* Full patrol reports should be produced and debriefed after each patrol action.

(4) *Tracking Presentations.* Full use should be made of tracking presentations when feasible. Each VT should report each of the incidents that he located.

0608. **Observation and The Use of Other Senses.** A high standard of basic infantry skills and fieldcraft is essential to the VT. His observation, hearing and sense of smell must be of the highest order. He should, therefore, utilise the following practices in order to hone his senses.

a. *Observation Training.* The VT should frequently practise observation training as follows:

(1) *Static Observation Stands.* Static observation stands help the tracker to practice scanning and searching both of which are essential at incident sites and when carrying out the track pursuit drill.

(2) *Mobile Observation Lanes.* These lanes are utilised for normal patrol training purposes, i.e., lead scout training.

(3) *Booby Trap Lanes.* Similar to mobile observation lanes except booby trap markers are used. These lanes are designed to promote booby trap awareness.

b. *The Sense of Smell.* It must be remembered that the majority of Vietcong camps in Vietnam were identified not through sight but through smell, as most lead scouts were able to detect the odour of both enemy latrines and cooking between 200 – 500 metres from the camp. In contrast the average distance of locating an enemy bunker or trench visually was between 3 – 25 metres. Trackers should, therefore, regularly track scent lanes in order to improve their sense of smell. Scent lanes are set up in a similar fashion to mobile observation lanes. However, various scents are laid on track to be detected by the tracker. Vegetables, i.e., garlic, onion, herbs and spices, perfumes such as soaps, balms and ointments, medicines, latrines and fires should all be used.

c. *Memory Training.* Memory retention is a vital attribute to the VT and he should regularly practise memory retention exercises. Examples of memory retention exercises can be found at Annex A to this chapter.

0609. **Tracking Techniques and Procedures.** The tracker must be taught and thoroughly practised in the fundamentals before he moves on to learning the tracking techniques and procedures. He should be taught in the following manner:

a. *Track Pursuit Drill (TPD).* In order to conduct the TPD the VT must have mastered pace to pace tracking and confirmed that he can recognise and pursue sign. In order to conduct the TPD the track must be laid by more than one enemy, two to four enemy are ideal. The tracker then pursues sign whilst conducting the TPD. The TPD can also be practised in pairs with the VT handing over to the CM every couple of bounds. This is useful in small areas or when few instructors are available.

b. *Track Casting Drill.* This is best taught as a lecture demonstration. It is practised as follows:

(1) *Initial Practice.* Track casting should be introduced during the TPD training and can be utilised when either the sign is lost or when the instructor deems necessary. Initial probes and initial casts only should be used.

(2) *Further Training.* It is inevitable that the tracker will have to conduct the track casting drill on a number of occasions during his training. As the VT becomes more competent the instructor should ensure that the VT conducts both extended casts and likely area searches in order for the VT to master these skills.

c. *Duties of a Coverman.* The duties of a CM can only be practised during pairs or team tracking exercises. These skills are best honed as follows:

(1) *Initial Training.* As previously mentioned the TPD may often be conducted in pairs. If this is the case the second VT should act as CM as a means of practice.

(2) *Advanced Training.* Tracker on tracker exercises are an excellent means of improving the skills of both the lead scout and CM. The layers will patrol out a certain distance then go firm in a fire position. The VT and CM will track them and whoever sees the other first engages with blank fire. This is a very good competitive means of developing the patrolling skills of the VT and CM and can be developed further by the laying party breaking track and setting ambushes, etc. A debrief should be held after each engagement.

d. *Incident Tracking.* Incident tracking should start from very simple incidents and should progress on to major incidents where a lot of information can be gained. Incident training must be progressive and should be practised as follows:

(1) *Incident Stands.* An incident stand is a fenced area that contains sign that an incident has occurred. They are used during the early stages of training to practise the VT in looking for sign, information and recognising incidents without worrying about the tactics. Initially the incident stand should include one or two mens' sign and eventually they should be enlarged to include the sign of three to four men. Once the VT demonstrates that he can gain all of the information available from such sites he should move on to tracking simple incidents.

(2) *Simple Incidents.* Simple incidents containing four to five mens' sign are the next step. The incident sites should not be fenced off or marked in any way other than the sign actually left behind. The purpose of these incident sites is to practise the trackers in both finding sign and the procedures of tracking an incident.

(3) *Detailed Incidents.* Detailed incidents are designed to practise the tracker in gaining information from the track. He should apply all of the correct procedures but whilst tracking should attempt to build a complete picture as to what has happened at the incident site. For example incidents may range from a short-term halt location occupied using a fishhook, snap ambush and then all round defence with a sentry deployed, to full scale platoon ambushes which either have or have not been sprung.

e. *Track Isolation.* Trackers when competent should be encouraged to carry out track isolation from time to time in order for them to gain confidence in the drill. In addition the instructor may use this drill during the early stages of tracking when tracking is too difficult due to fouling, etc.

f. *Deception Tactics.* Once the VT is competent at pursuing sign he should be taught deception tactics. This can be done by a lecture followed by tracking a number of deception lanes. Each lane should be laid by at least four men and should include a variety of deception tactics. Once the VT can recognise deception tactics deception measures should then be used on track at frequent intervals.

Summary

0610. The importance of a good instructor to student ratio during the initial stages of training the VT cannot be over emphasised. To teach the VT the lessons listed above and practise him accordingly will take between 10 – 14 days. The instructor will have to be both patient and diligent. Some soldiers will take to tracking far more readily than others but it is rare that a soldier will not be able to recognise sign at all. The major hurdle is getting the VT to have confidence in his ability. The instructor must always be ready to offer assistance and encouragement when required. Proficiency at visual tracking can only be achieved by frequent practice and closely supervised training.

0611 – 0612. *Reserved.*

SECTION 3. — TRAINING THE TRACKER TEAM

Introduction

0613. The TT must train until all members are competent trackers, and until they are fully functional as a team. The training of the team is initially the responsibility of the tracking instructor. However, once trained it is the responsibility of the team commander. Much of the training will be further practise of the skills listed above by tactically tracking an enemy over a period of days as a team with VTs rotating every couple of hours.

Initial Training Requirements

0614. **Routine Drills.** The team will utilise the routine drills contained in Chapter 4 whilst tracking. It is essential that the team becomes fully proficient at conducting each drill in order to ensure that by negating the requirement for mundane orders the majority of time is spent on track. Tactical halts as well as overnight/LUP halts should be practised thoroughly before the first tracking exercises.

0615. **Contact Drills.** The team must practise contact drills from all sides and in all types of terrain until the drills become second nature and due to the rotation of the VT each man must be capable of applying the drill regardless of where he might find himself in the order of march. During training the team must be exposed to contacts, including booby traps, so that they maintain a high standard of alertness throughout the patrol.

0616. **Methods of Insertion.** The team must be trained so that it is capable of insertion by vehicle, boat or helicopter. Vehicle and boat insertion will be utilised where access is simplest by those means. Helicopters will often be used in close country and where LPs don't exist the team must be trained in helicopter abseiling and fast roping.

0617. **Tracking Exercises.** The team should undertake supervised tracking exercises with as much time allocated for battle procedure as possible. The exercises should be progressive in both difficulty and duration. Early tracking exercises should be short and because the speed of tracking is likely to be slow it is normal that only the VT and CM are 100% alert at all times. This will change as and when tracking speed increases. The age of sign should be no more than 24 hours old for initial exercises, reaching 72 hours old on the later exercises. In addition the enemy tactics used should be simple early on with larger amounts of sign being left in order to get the team used to looking for sign. On completion of each exercise the team should be fully debriefed and their findings compared to the details provided by the laying patrols.

0618. **The Pursuit.** TTs should be trained in pursuit including the use of dogs where available. Pursuit exercises must include a full brief on arrival followed by a pursuit and follow-up action. Pursuit exercises should be introduced during the later stages of training when the VTs can easily recognise sign and are capable of tracking quickly. The aim of pursuit training is to get the team tracking faster than the enemy is moving.

Continuation Training

0619. Continuation training, even during operational deployments, is essential in order to prevent skills fade. The training should include the following:

    a. *Basic Patrol Training.* Normal patrol training is carried out with the emphasis on tracking routine drills and procedures. This will also include practising the following skills:

        (1) Navigation.

        (2) LP and Winch hole selection including ground to air signals.

        (3) CASEVAC Procedures.

        (4) Health and Hygiene.

        (5) First aid training.

b. *Visual Tracking and Observation.* Observation is continually tested during visual tracking exercises. However, observation tests should be conducted weekly in the form of mobile and static observation lanes. This training is given to the entire team to assist them in obtaining additional information whilst following the enemy.

c. *Contact Drills.* Contact drills should be constantly rehearsed including dry, blank and live firing. Each man should regularly practise being a lead scout by utilising snap shooting ranges and individual CQB lanes.

d. *Physical Training.* It is vital that the team is physically fit. This is achieved by scent and visual tracking exercises and also by regular PT. The time a team can spend on task will be determined not only by its tracking ability, but also as a direct reflection of its fitness. The fitness of the handler and especially his dog must not be overlooked.

e. *Helicopter Abseiling.* The basic drills need to be rehearsed and the use of abseiling equipment practised as often as possible. This includes abseiling with dogs.

f. *Signals Training.* Every member of the team must be proficient in the use of VHF and HF sets, and proficient at sending the various signals reports.

0620. **Tracking Related Subjects.** The TT should also be skilled at conducting the following actions:

a. *Ambush Reconnaissance.* This is a particularly useful skill when a MSR or a route which is subjected to significant use is found. The TT should be tasked as a matter of course to conduct an ambush reconnaissance.

b. *Close Target Reconnaissance (CTR).* This is a vital skill especially when an enemy camp is found. In order to ensure that time is not wasted the TT may conduct the CTR of the enemy camp.

c. *RV Procedure.* RV procedure is essential when linking up with other friendly forces, for example when leading them into a deliberate attack or into an ambush site.

Summary

0621. The training received by the TT during their initial training is the minimum required in order to be able to attempt the task of tracking the enemy. A diligent, robust approach to training must be maintained if the team is to reach its full potential as an intelligence gathering asset.

0622 – 0623. *Reserved.*

RESTRICTED

SECTION 4. — TRACK AND INCIDENT LAYING

Introduction

0624. It is important that all tracks and incidents are laid correctly by the laying patrols in order to ensure that the tracking exercise runs as the exercise director has planned. Therefore it is vital that the laying patrol commander and his patrol are diligent when both conducting and recording their actions.

0625. **Briefings.** The laying patrol commander must be briefed comprehensively by the exercise director/tracking instructor in order to establish exactly what he is required to do. The laying patrol commander should ideally be tracker trained. However, as a minimum must have some knowledge of tracking techniques. The tracking instructor should give the laying patrol commander a patrol trace illustrating the route to be taken and should tell him exactly what actions he is to carry out on route. This briefing will include the distances between incidents and all SOPs he is to employ if the VT is to obtain the intended track picture. The instructor's brief will include the following:

    a. Formations and order of march to be used by the laying patrol.

    b. Patrol routine drills/SOPs to be used.

    c. Details of any discardables to be left at each incident site.

    d. What equipment is to be carried/worn including individual places within the order of march, i.e., the radio operator, types of weapons to be carried, foot wear, etc.

0626. **Laying Patrol Options.** Once the laying patrol commander has obtained his brief from the exercise director and completed his estimate, he will then brief the patrol as to what is required of them. This can be done in two ways and is dictated by the exercise director as follows:

    a. *Normal Orders.* Firstly it could be run as an exercise so that the troops laying the track conduct their normal unit SOP as they would during the course of any normal training exercise. Therefore the brief would be given as a formal set of orders. This first method can benefit the laying patrol and the TT but it is difficult for the commander to accurately record all the information.

    b. *Set Patrol Actions.* The second and the preferred method for training TTs is choreographing the laying patrol action allowing all the information to be easily recorded by the laying patrol commander. The benefit here is not only the reduction in time taken to record the incident, but when the TT debriefs on the information gained the laying patrol commander can readily confirm the accuracy of that information.

RESTRICTED

RESTRICTED

0627. ***Recording the Laying Patrol Action.*** The laying team commander will use a track laying report (*see* Annex B) to record his actions at each incident site. This report must be compiled accurately as it will be compared directly with the report subsequently compiled by the TT. Each incident conducted by the laying patrol must be sketched (*see* Appendix 1 to Annex B).

Summary

0628. The success of each tracking exercise is largely dependent on the laying patrol conducting its action in a professional and diligent manner. It is essential that the sign is not over emphasised, i.e., deliberate kicking of tree roots, dropping of radios and weapons, etc., to create better sign.

0629 – 0630. *Reserved.*

SECTION 5. — TRACKING PRESENTATIONS

Introduction

0631. Tracking presentations are a useful tool when training both the VT and TT. Due to the potential use of this type of brief TTs should be well practised in the preparation and delivery of these presentations. The presentation is delivered as a formal activity and should be attended by as many tracker trained personnel as possible. The laying patrol commander must attend as should the person who gave the patrol its initial orders.

Method

0632. The patrol will deliver the presentation using the method and aids described in Chapter 5. Questions should be asked where the information provided by the TT is either unclear or incorrect. At the end of the presentation the laying patrol commander should present each incident he has laid and either confirm or deny the information gained by the TT. For the purposes of the exercise each incident encountered should be presented. Following the presentation the tracking instructor should summarise the overall effectiveness of the team's action.

Summary

0633. The tracking presentation is an excellent training tool as it allows the TT to make a direct comparison of the information they have gained with the information available. It helps the team to build confidence and is an invaluable training assessment tool.

RESTRICTED

ANNEX A TO
CHAPTER 6

MEMORY TRAINING

Introduction

1. Memory is the process of storing and retrieving information in the brain. To 'remember' something is to recall a past experience to present consciousness. This may be either a personal experience or acquired knowledge. Memory is based on each individual's powers of retention. The efficiency of each individual depends upon how accurate and lasting his powers of retention are, and upon his ease of recalling that information. This Annex is designed to demonstrate some methods of remembering things in order to aid recollection and recognition which will be of immense value to prospective trackers.

Types of Memory

2. Psychologists ordinarily discuss four different types of remembering:

   a. *Recollection.* Recollection involves the reconstruction of events or facts on the basis of partial clues which serve as reminders, e.g., music, occasions, i.e., hotel safe numbers/birthdays, etc.

   b. *Recall.* Recall is the active and unaided remembering of something from the past of something you have been involved in.

   c. *Recognition.* Recognition refers to the ability to correctly identify previously encountered stimuli as being familiar, i.e., recognising a face, roundabout or pub on a corner.

   d. *Re-learning.* Re-learning is showing evidence of the effects of memory. Material that is familiar is often easier to learn a second time than it would be if it were unfamiliar.

3. Recall and recognition usually go together but when, for instance, you know the face but are unable to put a name to it, that is an example of recall occurring without recognition.

Long Term and Short Term Memory

4. There are numerous methods utilised to improve long term memory. Long term memory is an excellent asset to the VT. The most popular methods are described in subsequent paragraphs.

RESTRICTED

5. **Method 1 — Mnemonics.** Remembering by mnemonics is both a common and simple practice encountered in all types of military training. A list is remembered by taking a prominent letter from each item in the list, usually the first letter, and forming a word with those letters as follows:

   a. *Example 1.* The sequence of a fire control order is remembered by the mnemonic GRIT meaning:

   G — Group
   R — Range
   I — Indication
   T — Type of fire

   b. *Example 2.* The principles of defence are remembered by the mnemonic DAMROD meaning:

   D — Depth
   A — All round defence
   M — Mutual support
   R — Reserves
   O — Offensive spirit
   D — Deception and concealment

6. **Method 2 — Association.** In a purely military setting this method can be easy to master. Because we know certain facts, i.e., soldiers wear uniforms, wear equipment, carry weapons, etc., we can concentrate on other detail, i.e., types of uniform, colours, headgear, webbing, specific weapon types, etc.

7. **Method 3 — The Peg or Hook System.** The peg or hook system allows you to remember words in any order. It also works by using imagery to link up a new word to an image.

   a. Before you can use this system you have to learn a very simple poem for example:

   'One is Bun, Two is Shoe, Three is Tree, Four is Door, Five is Hive, Six is Sticks, Seven is Heaven, Eight is Gate, Nine is Wine, Ten is Hen.'

   b. To use the peg system picture the first new word you want to learn with the first new word in the poem, the second new word with the second word in the poem and so on. If the first three new words to learn were Dog, Plate and Arrow, you would picture a Dog eating a Bun (first peg word), a Shoe on a Plate (second peg word), and an Arrow hitting a Tree (third peg word).

8. **Method 4 — Extempore/Story Line.** This method requires the observer to make up a story line whilst viewing the scene or objects. The more bizarre the story the easier it is to recall, for example:

RESTRICTED

Photo	I was looking at a Photo
Flower	of a Flower. There was
Dog	a Dog eating the flower. It washed it down with
Beer Can	a Beer, then climbed on top of an Elephant.
Elephant	The Elephant was eating
Apple	an Apple whilst reading
Book	a Book and watching
Video	a Video. The Video showed a car driving too fast
Tyre	for the bend and the Tyre blew out.
Gun	It made a sound like a Gun.

9. **Method 5 — The Story Method.** You do not need to use the peg method in order to use imagery. You can simply link all the words you want to remember together in pictures, one after the other. Suppose the words you want to remember are Dog, Car, House, Snow and so on. First, you picture say a Dog chasing a Car. Then you picture a Car crashing into a House. Then you picture a House covered in Snow and so on. Each word automatically leads to the next one. It is a highly effective method of remembering, provided you make vivid pictures to link the words together, but if you forget one link, the rest of the words are likely to be forgotten.

Summary

10. These are just a few of the many methods available to assist a tracker with memory retention. You can improve the amount of material retained by practising 'active recall' during learning, by periodic reviews of the material and by over learning the material beyond the point of bare mastery. Basically you must train your brain by conducting regular memory training.

RESTRICTED

RESTRICTED

**ANNEX B TO
CHAPTER 6**

TRACK LAYING REPORT

Callsign:

Area of Operation:

Map References: Northings: Eastings:

Patrol Composition:

    a.

    b.

    c.

    d.

    e.

    f.

Equipment:

Weapons:

Mission:

RESTRICTED

RESTRICTED

Timings:

    a. Infil: ---

    b. Exfil: ---

Infil point: Grid: ---------------------------- Method: -----------------------------

Exfil point: Grid: ---------------------------- Method: -----------------------------

Ground in general:

Ground in detail:

Climatic Conditions:

Day 1:

Day 2:

Day 3:

Day 4:

6B-2   **RESTRICTED**

RESTRICTED

Route taken:

Day 1:

Day 2:

Day 3:

Day 4:

Patrol Commander's Sig

RESTRICTED　　　　　6B-3

RESTRICTED

RESTRICTED

RESTRICTED

**APPENDIX 1 TO
ANNEX B
TO CHAPTER 6**

Incident No:

D.T.G:

N

Grid:

Key:

RESTRICTED

RESTRICTED

RESTRICTED

www.ingramcontent.com/pod-product-compliance
Lightning Source LLC
Chambersburg PA
CBHW070321190526
45169CB00005B/1692